Advance Praise for

microhydro

At long last, a book about small, residential-scale hydroelectric.
Our industry has been looking for this book for over ten years!

— DOUG PRATT, Technical Editor, Real Goods

As manufacturers and marketers of the PowerPal line
of microhydro products, we are really delighted with Scott Davis'
new book *Microhydro: Clean Power from Water*. It is a well-written,
informative and, above all, practical guide to all aspects
of microhydroelectric generation and installation.
We certainly intend to recommend it to our
customers and to other interested parties.

— DAVID L. SEYMOUR, President, Asian Phoenix Resources Ltd.

This book reflects Scott Davis' many years of experience
in this field, and his belief in energy conservation.
He presents the art of small micro-hydropower generation
in a very easy to read format, with excellent illustrations.
This book would be invaluable to anyone planning
to develop a small microhydropower system and
to all who believe in renewable energy.

— GHANASHYAM RANJITKAR, Energy Specialist
(Micro-hydropower), MIEE

Finally! A book that explains how microhydro works.
Scott has filled the long-standing need for this kind of book
with one that is comprehensive and understandable.

— POWERHOUSE PAUL, manufacturer of micro-hydroelectric
generators since 1980

microhydro

microhydro

Clean Power
from Water

SCOTT DAVIS

NEW SOCIETY PUBLISHERS

Cataloguing in Publication Data:
A catalog record for this publication is available from the National Library of Canada.

Copyright © 2003 by Scott Davis.
All rights reserved.

Cover design by Diane McIntosh. Water image ©Photodisc.
Printed in Canada. Fourth printing May 2008.

New Society Publishers acknowledges the support of the Government of Canada through the Book Publishing Industry Development Program (BPIDP) for our publishing activities.

Paperback ISBN: 978-0-86571-484-7

Inquiries regarding requests to reprint all or part of *Microhydro* should be addressed to New Society Publishers at the address below.

To order directly from the publishers, please call toll-free (North America) 1-800-567-6772, or order online at www.newsociety.com

Any other inquiries can be directed by mail to:

New Society Publishers
P.O. Box 189, Gabriola Island, BC V0R 1X0, Canada
1 (800) 567-6772

New Society Publishers' mission is to publish books that contribute in fundamental ways to building an ecologically sustainable and just society, and to do so with the least possible impact on the environment, in a manner that models this vision. We are committed to doing this not just through education, but through action. We are acting on our commitment to the world's remaining ancient forests by phasing out our paper supply from ancient forests worldwide. This book is one step towards ending global deforestation and climate change. It is printed on acid-free paper that is 100% old growth forest-free (100% post-consumer recycled), processed chlorine free, and printed with vegetable based, low VOC inks. For further information, or to browse our full list of books and purchase securely, visit our website at: www.newsociety.com

NEW SOCIETY PUBLISHERS   www.newsociety.com

Books for Wiser Living from Mother Earth News

Today, more than ever before, our society is seeking ways to live more conscientiously. To help bring you the very best inspiration and information about greener, more sustainable lifestyles, New Society Publishers has joined forces with *Mother Earth News*. For more than 30 years, *Mother Earth News* has been North America's "Original Guide to Living Wisely," creating books and magazines for people with a passion for self-reliance and a desire to live in harmony with nature. Across the countryside and in our cities, New Society Publishers and *Mother Earth News* are leading the way to a wiser, more sustainable world.

I'd like to dedicate this book to my family,
who have done so much over the years
to make it all possible.

Contents

Acknowledgments

M Y PARTNER AND SPOUSE, Bonnie Mae, has gotten me out of a lot of trouble, from ditch digging to "agreement in number and tense." Our daughter, Alannah, learned her multiplication tables walking up and down a mile long penstock under construction, and decades later, reviewed this book in manuscript. My mother, Della, made sure that I was literate. My father, Roy, although he knows how things should be done properly, was never afraid to make temporary repairs on very old equipment – repairs that lasted for years.

Bob Mathews of Appropriate Energy Systems was always the best microhydro guru…

I would also like to acknowledge those who have helped with the book project.

Dr. Michael Danov and Alannah New-Small reviewed the draft at an early stage.

Thanks also to Sheila Potter, Ghanashyam Ranjitkar, Eric Smiley, David Seymour and Paul Cunningham.

Any remaining errors are of course my own.

And finally, many thanks for the opportunity to tell a few stories to the whole New Society crew, including Chris and Judith Plant, my editor Ingrid Witvoet, and the long suffering Greg Green and Jeremy Drought.

Introduction

GENERATING ELECTRICITY FROM WATER POWER, HYDROELECTRICITY, is the largest source of renewable energy in the world today.

Microhydroelectric systems generate electricity from small water powered alternators.

Even at the smallest of scales, water power continues to be a most reliable and cost effective way to generate electrical power with renewable technology. Yet, getting the relevant information to recognize and develop a microhydro site successfully has been remarkably difficult to find for many years.

I dropped out of graduate school in 1977 to work on a project that included, among other things, a village scale microhydro project. In the decades that followed, I owned, operated, repaired, sold and generally fooled around with microhydro technology. Every site I visited had many unique features and some common ones as well. Out of this experience, this book brings you a range of solutions to supplying energy from flowing water, from the smallest and simplest systems, up to a relatively high output system.

There's more than one way to read this book. If you want to know about the topic as a whole, just start at the beginning and go all the way through. You will get an idea of the range of sites that are practical to develop, how different site features can be opportunities or obstacles, and something about costs as well.

Or, you may have a particular site in mind, and a pressing need to figure out what to do. Let's say it's getting on to be winter and you are wondering if you could develop a renewable energy supply that, unlike solar, would work through the dark part of the year.

Here's how you can get what you need from this book:

It begins with a short introduction to electricity and to hydraulics, which describes how water behaves. If you are confident in these areas, you can skip this part.

But then, do the self assessment, Chapter Two, to understand what size and kind of system will meet your needs. If you believe that you have a clear idea of your needs, and the budget to make it happen, you could skip this part.

Next, move to assessing your site, Chapter Three, to see if you can find the waterpower potential you need. If your first survey turns up empty, go back to the self-assessment and see if a smaller system could meet your needs with electronics rather than with waterpower alone. Many small sites go unappreciated.

After you assess your needs and your site, everyone should read Chapter Four. It will assist you in determining the system that is appropriate for you by comparing how you rated your needs in the self assessment score with systems that have met those needs successfully. It explains the various technologies that are brought together to make a successful microhydro system, from small battery charging systems to ones that are large enough for a small village. If you are in a real hurry, just read about the kind of system you are planning. However, it never hurts to be familiar with the alternatives that are available to you.

Chapter Five goes into detail about getting started with your microhydro system. Everyone should read this chapter as well.

Chapter Six discusses incentives and regulations. These are important, local factors that can easily make or break a project.

Chapter Seven is a collection of case studies. Here you can find examples of successful systems, one of which probably looks like the one you are planning. Each illustrates important elements in successful microhydro developments.

There's a glossary at the end to help with any unfamiliar terms.

Now you're ready for Chapter One, What is Microhydro?

> **Many times, people overestimate the amount of power they need. This may lead them to overlook significant resources.**

What is Microhydro?

Introducing Electricity and Hydraulics

T HIS BOOK IS ABOUT MAKING ELECTRICITY FROM WATER POWER. Thus you
do need to know something about how electricity works, and a bit about
how water behaves. Luckily, the engineering has been done long ago, and
so you just need to know enough to make informed choices.

If you feel you already have a grasp of the fundamentals, skip this part and
go directly to the next chapter on assessing your site and your requirements.

About DC

Direct current (DC) was the original kind of electricity. Although alternating
current (AC) is the most commonly used form of electricity today, DC is
still found everywhere in modern life. Any
time power is to be stored, or used without
a connection to the power grid, chances are
that it is DC.

Batteries are the most common source of
DC. A battery is like a storage tank of
electricity. The power comes out one side or
pole of the battery and flows to the other
side. These poles are called positive and
negative. Your watch, your laptop, the
starter on your car, and a multitude of other
technologies in everyday life use DC.

A Turgo microhydro turbine. Credit: Ann Cavanagh.

There are other ways in which DC is different from the plug-in AC power that we are all used to as well. Most DC applications are lower voltage, such as 12 or 24 volts, while AC voltages in common use are 120 or 240 volts, or even higher.

DC has many uses, but it is not the same as the power that comes out of your plug-ins. Many things that you want to plug in cannot be plugged directly into batteries.

There are many special appliances adapted for DC. However, the best and cheapest equipment runs on 120-volt AC sine wave current — just like the standard North American current.

About AC

AC is the kind of electricity that comes out of your wall outlets. AC means "alternating current," which means that it changes direction. An important feature about changing direction is how fast it changes direction, which is measured as "frequency." North American power has a frequency of 60 cycles per second. In some other parts of the world, the AC frequency may be 50 cycles per second.

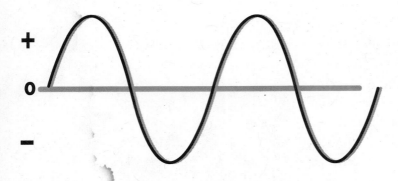

AC voltages are higher than DC voltages. Electricity comes out of outlets in North America at 120 volts or thereabouts, and 240-volt circuits are used in homes for large loads such as heaters.

The process of changing DC into AC is called "inverting" and the device that does it is called an "inverter." Converting 120-volt AC into 12-volt DC is a

The sine wave graphs the way alternating current changes direction. Credit: Corri Loschuck.

relatively easy process electronically. The transformer to bring the AC to 12 volts and the rectifiers to make the AC into DC by flowing only one way is old technology, and has been widely used for years. However, changing battery power into AC that looks like it came from an outlet is a real trick.

One of the reasons that microhydro isn't everywhere is that inverters were only perfected in the late 1980s. They were available earlier, but they were unreliable and expensive. Since they were expensive, they were often on the small side for the jobs asked of them. As a result, they were routinely overloaded, and their reliability suffered.

Moving Power

Power needs to get from where it is generated to where it is used, and losses due to resistance are inevitable. As with frictional losses in pipes, which can be corrected through use of larger pipes, electrical losses are reduced by using a larger conductor. And in both cases, a cost effective solution balances losses and cost.

Electricity and Water

If you keep your sense of humor about it, electricity is "like" water, in many ways. For example, the flow of water is like electrical current. Where the flow of water is measured in gallons per minute, electrical current is measured in "amperes" or "amps" for short. Amperes are abbreviated as "A".

Now to carry the analogy further, consider water pressure. Water pressure is "like" electrical voltage. The unit of voltage is the volt (V). The amount of pressure you read on your gauge is also a direct measure of the height or head of water above it. Pressure is measured in pounds per square inch and is described as "feet of head."

The pressure in a pipe is highest when no water is flowing. This is called the static pressure or static head. As water begins to flow in a pipe, friction takes its toll and some of the pressure is lost to friction. The more water flows, the more pressure is lost. Since pressure and head are the same, pressure loss can also legitimately be called "head loss." The pressure that remains from the static head when frictional losses are subtracted is called the "net" pressure or net head.

Electrical resistance, measured in ohms, works very much the same way as pipe friction. As current flows, potential is lost to resistance. The more current flows, the more is lost. This is called "voltage drop."

Calculating Electrical Characteristics

Current, voltage, and resistance are all related to each other in a pretty simple way — current equals voltage divided by resistance. Thus, you can calculate any value, as long as you have the other two. For example, voltage equals the current times the resistance.

Water power is the product of flow and pressure, and electrical power is the product of amperage and voltage. Just as horsepower would be the unit of shaft power, the watt (W) is the unit of electrical power. A kilowatt is 1,000 watts. We pay our electrical bills by the kilowatt hour, which is 1,000 watts used for an hour.

Hydroelectricity

Water power has been used to power equipment for tasks such as milling grain and pumping water for many hundreds of years. Slow moving waterwheels are ideal for some kinds of jobs, and there are many traditional designs.

Generating electricity is the kind of job for which traditional waterwheels are less suitable. Generally, making electricity requires rotating machinery at many hundreds of rpm, while water wheels may typically have rotational rates of a dozen rpm or less. During the nineteenth century, the traditional waterwheel was made more efficient. The Poncelet wheel, with a vertical spindle and a runner with a curved blade, evolved into the Francis turbine which is in very common use today as a hydroelectric turbine.

In 1866, a ten-year-old Nikola Tesla had a "vision" of harnessing the power of Niagara Falls with "some kind of wheel." In the 1880s, the first hydroelectric power systems were developed.

Focus on Microhydro

Here, the focus is on sites that are big enough to power a household, or a few households, in the North American manner. In order to provide a bit of perspective on just how much power North Americans use, we have also included a case study from a remote village in the Philippines.

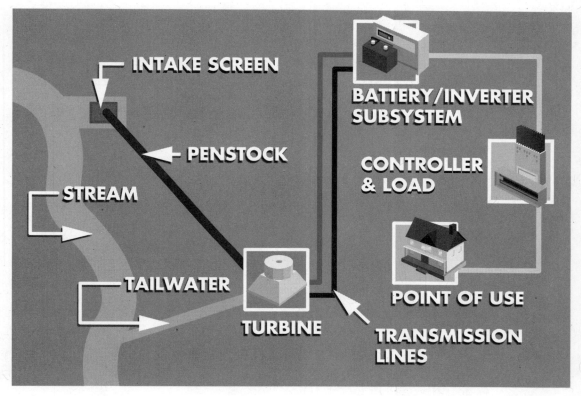

A typical system. Credit: Corri Loschuck.

Microhydro uses the same kinds of turbines which are used in larger systems. Improvements in microhydro practice have not been driven by innovations in turbine design, but rather by improvements in two areas of balance of system components — electronic load controllers for AC systems, and inverters for battery charging systems.

Prior to the 1980's, the flow of water was controlled to keep the alternator speed steady by activating a needle nozzle. This combination worked well, but was expensive to purchase and operate; the arrival of electronic load controlling meant that power was used where required. In a typical household, if a light is turned off, then the hot water tank or space heating is exactly that much hotter; all power produced goes somewhere useful. The development of this technology has contributed significantly to the spread of microhydro technology.

Additionally, battery charging systems and small sources of power in general were hampered by early inverters, which were expensive, fragile, noisy acoustically and electrically, and not very powerful. Dramatic improvements in modern inverters means that the number of sites that are practical to develop have increased accordingly, considering that there are many more sites that can produce a few dozen or hundred watts of DC charging power than those that can produce a few kilowatts of AC.

Microhydro technology will likely continue to improve over time. As the technology advances and more efficient appliances and light sources continue to be developed, microhydro promises to continue to offer more for less.

In a typical system, water leaves a stream through an intake ditch or canal. Water is brought to the forebay, where intake screens remove debris that could clog the jets of the turbines. Water flows from the intake through the penstock to the turbine and generator unit, where power is generated. The power is transmitted to a battery/inverter subsystem near the point of use.

Good News, Bad News

As explained, microhydro is a reliable, well worked out technology, but it does have advantages and disadvantages.

First, the good news: If you have a site, microhydro can deliver the best bang per buck when it comes to providing electrical service from renewable energy. Significant power can be generated with flows of two gallons per minute, or from drops as small as two feet, and power can be delivered, in a cost effective fashion, a mile or more from where it is generated to where it is being used.

Microhydro power is continuous power, unlike the sun, which goes down at night, south for the winter, and can go behind a cloud at any time; or the wind, which blows at some times and not others. Sites can easily be winterized to provide for the winter months when the solar resource may be almost non-existent.

Additionally, existing pipelines, such as those delivering gravity domestic water to a house, can often be used as penstocks to generate electricity.

Taking all this into account, a small microhydro system can be developed for as little as $650 (Cdn$1,000). With the help of this book, you will be able to do the much of the installation yourself (most microhydro systems are owner installed), which naturally contributes to the low cost.

If you are connected to the utility grid, in some Canadian provinces and in 35 American states, you can either offset your own electrical consumption (net metering) or be an independent power-producer without too many barriers.

There are, however, some disadvantages to microhydro.

Although low cost microhydro is possible, certain flow, head, and output characteristics are required. More conventional systems in more usual circumstances require considerable balance of system components, including pipeline, transmission line, controller, batteries, and inverter. A more typical whole system cost would be about $1,900–$9,700 (Cdn$3000 –$15,000), about the same as a motorcycle or a used pickup truck.

Not all sites where there is potential energy available will allow microhydro to be developed in a cost-effective fashion. Particularly, extracting multiple kilowatts from heads under about 50 feet requires turbines that are too large and expensive for general use.

Regardless of output, there are certain fixed costs. For very small loads, like lights that are used occasionally, or small equipment loads, photovoltaics may be more economical because you only pay for the capacity that you use.

Pipelines for microhydro can be thousands of feet long and can easily be the most expensive and difficult part of the project. You have to know when to quit.

We use the word "optimum" to mean that it's time to quit. For example, the optimum flow in a pipe is considered to be

A likely stream. Credit: Scott Davis.

the point where running more water through the pipe won't make any more power. Then you should quit.

Of course, it is also quite true — in water power as well as in other areas of life — that it might be smart to quit while you're ahead. For example, while it is true that a certain flow in a certain pipe will produce the most power, it is also true that diminishing returns are setting in, and using somewhat less water in the same pipe makes but a small difference in the power output.

It may be that taking the trouble to handle more water isn't worth the bit of extra power you get. (See the case study, "Upgrading a Small AC System.") Then again, you might want to run a lot of water through the pipe for frost protection. (See the case study, "A Simple System.")

Technically, microhydro outputs range up to 100 kilowatts, but here we are most interested in the lowest part of the scale. After all, the smaller the output, the larger the number of sites. Even the tiniest amount of power is better than nothing.

If you have a site to develop, or think that you might, this book will equip you with a good overall view of the topic. With a clear idea of your needs, you should be able to assess a potential site to see if it is appropriate.

You will get to know the relative strengths and weaknesses of not only microhydro, but also of wind and solar technologies which may also be appropriate at your site. You will be able then to see which technology or combination is right for you.

Finally, you will finish this book with a better idea of when to quit. Many times, people overestimate the amount of power they need. This may lead them to overlook significant resources.

In any case, welcome!

2

Assessing Your Energy Needs

W E LIVE IN A WASTEFUL SOCIETY. We may, by habit, use far more power than is necessary to do the things we do. After all, low energy prices discourage conservation. We need to look at consumption patterns in other countries to improve our perspective.

Off the utility grid, you are no longer the beneficiary of subsidized energy. Suddenly, you must develop the kind of awareness of the energy you use that is common in other parts of the world. And one of the first things you will see is that efficient appliances can do more with less. Just ask anyone who uses photovoltaic technology about conservation. Efficient lighting is at the top of the list of opportunities, then refrigeration.

In this chapter, you will find out how much power is required to meet your needs. Solar, wind, and microhydro systems are compared by their abilities to meet these needs.

The Kilowatt Hour

A confusing fact about different renewable energy technologies is that they measure their outputs in different ways, and it is hard to compare them directly as a result.

To solve this problem, you can compare the output of different technologies directly by the "kilowatt hours per month" (kWh/month) that they produce; after all, you've paid your electrical bills this way for

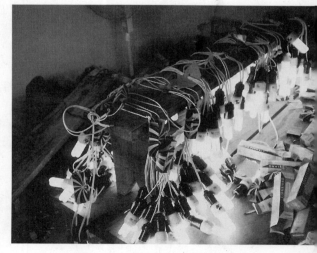

Even a small turbine will power a lot of energy efficient lights. Credit: Asian Phoenix Resources, Ltd.

11

years. Most of us have always bought our power from a utility and were billed for kilowatt hours consumed per month, and so this way of measuring power consumption at least is familiar to many.

However, buying power from a utility at subsidized prices may also confuse us by making us think that we need more power than we actually do for good service. Most people in other developed countries use far less power for their homes than we do in North America, and they don't seem to complain.

The service that you get from the power you have depends entirely upon the efficiency of the equipment you are using. North American appliances, from light bulbs to refrigerators, can be very inefficient. If you're exploring the idea of producing your own power, at your site, it may very well cost two or three times as much money and effort to produce two or three times the power, and so you will want to consider carefully the idea of efficiency.

Electrical Consumption and Capacity

When you generate your own power, you don't necessarily pay for the kilowatt hours you consume; rather, you pay for how much capacity you have. AC sites have to be big enough to start and run the largest motor required of the system, for example, as well as delivering the total number of kilowatt hours required per month.

Your power requirements change during the course of the day. In the middle of the night, you may not need any power. However, at random intervals during the day, your refrigerator and freezer will start and run for a while. They need a big surge of power to start, just for a second or two, and then they don't require much to run at all. When the sun goes down, lights go on and these are steady loads that are on for a few hours. Because your requirements are so variable, two kinds of power measurement are useful:

• Consumption, which is the average amount used.
• Capacity, which is the measure of the largest load required.

Comparing Service from Different Technologies

Many of us have pondered the question of solar power, since almost any home can be powered from the sunlight that falls on its roof. A microhydro site is likely to have solar and perhaps wind potential, too. The question is where to invest your energy budget most wisely.

TYPICAL POWER CONSUMPTION

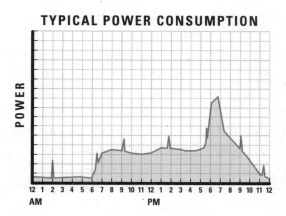

Your electrical needs vary over time.
Credit: Corri Loschuck.

PHOTOVOLTAIC

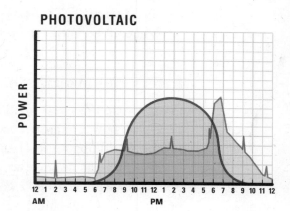

PV outputs drop to zero every night and are much reduced in winter and cloudy weather. Credit: Corri Loschuck.

WIND GENERATOR

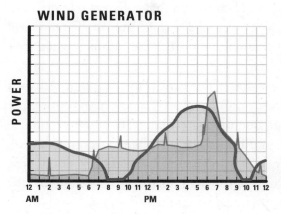

Wind system outputs go up and down irregularly. Credit: Corri Loschuck.

HYDRO

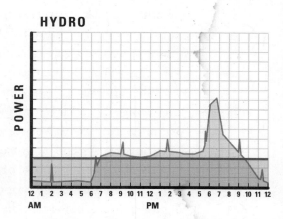

Microhydro systems have the same output all the time. Credit: Corri Loschuck.

One of our case studies, "A System With a Long Pipeline," produces 50 kilowatt hours per month or 70 continuous watts. Some useful questions are:

- What kind of service will this system provide?
- How does its service compare with a typical photovoltaic system rated at 300 peak watts?
- Where does that fit on our self-rating scale?

Rating Photovoltaic Outputs

Photovoltaic (PV) and wind generating systems are traditionally rated by their peak power output. This is the output you get from your system when full sun is shining on it, or the wind is blowing at a certain speed.

But the solar resource varies a lot. In order to tell how many kilowatt hours you will get from a PV system, you will have to know how many hours the sun will shine, adjusted for those hours when it is not shining quite enough for full output, in each month. This varies a lot from place to place. Summer is pretty nice everywhere. But the amount of sun you get in winter is quite different depending on your location.

In order to get an idea of how much service we get from a photovoltaic system, we pick an average month. Then we say that a system will deliver a certain number of kilowatt hours during this average month. Then, it will produce at least that much power and more during the summer, when days are longer and the weather is better.

For example, a PV system that has a 300 watt peak output will deliver an average of 42 kilowatt hours per month of power, and more during the summer. And so to answer our question, a 50 kilowatt hour per month microhydro system produces significantly more than a 300 watt peak photovoltaic system most of the year, and much more during the winter. Since lighting is a large proportion of the load for many small systems, the lighting load is highest in the winter, just when the photovoltaic resource is waning.

Winter weather varies a lot more by location than summer weather does, but expect photovoltaic system performance in December to be a tenth to a quarter of average.

Rating Wind Outputs

Similarly, the wind blows sometimes and not others, sometimes strongly, sometimes weakly. In order to tell how many kilowatt hours you will get from a wind machine, you need to know the average wind velocity at your site.

Rating Microhydro Outputs

Microhydro systems are rated by their peak output, but this is generally the same as the average output — they tend to continue to produce at their rated output all the time. As long as the water flows day and night, summer and winter, the system produces power. In short, the hydroelectric resource is not necessarily intermittent, like the sun or the wind, which means that a little bit can go a long way.

Microhydro systems run all day and all night, which is 720 hours in the average month. So, the continuous output (expressed as kilowatts) is multiplied by the 720 hours per month to get the kilowatt hours per month.

For example, take a hydroelectric system that puts out about 70 continuous watts, as in our case study "A System With A Long Pipeline," which makes about 0.07 kilowatts. Multiplying this times the 720 hours in a month will give you about 50 kilowatt hours per month.

> A microhydro system that puts out about 70 continuous watts will give you about 50 kilowatt hours per month.

If the peak output of a microhydro system is too small, a battery and inverter subsystem is used to give the peak output desired. Since microhydro systems often have power to spare in terms of kilowatt hours, the performance of a system is dictated much more by the size of the inverter than by other factors.

Different Jobs — Different Power Requirements

It is important to remember that you are in the business of providing yourself with service: for example, "lighting," rather than generating a certain amount of power.

Many loads are intermittent. For example, lighting goes on at dark and stays on until everyone goes to bed at night. Batteries can charge up the rest of the night and during the day.

Many important tasks, like motor starting, use quite a lot of power to start, and then a much more modest amount to run. It is entirely possible to have a system that has plenty of kilowatt hours available, but not enough capacity to start critical loads. I traded a fairly ordinary washing machine that I had fitted with an easy start kit, for my neighbor's nice new Maytag washer, because his inverter would not start the direct drive Maytag machine. It did just fine with the one we'd modified.

Assessing Your Requirements, Assessing Your Site

The first part of a successful microhydro development is to find out what it is that you want to do. Satisfaction, not watts, is your goal.

Making a system that is neither too big nor too small is an important factor in successful projects. Large projects have a way of not getting done, and thinking that large projects are required for satisfaction is a common error.

Looking at your electrical bill may not give you a very good idea of how much power you actually need for good service. The low prices of North American energy have led people to use more power than they need. Market forces dictate that in Canada, for example, people use about two and a half times as much utility power as people do in Denmark.

> Looking at your electrical bill may not give you a very good idea of how much power you actually need.

Even if you are off-grid, and are using a fossil-fuel-powered generator, your fuel consumption won't necessarily reflect how much power you need either. Despite the fact that generating electricity with fossil fuels is quite expensive, conservation is often overlooked. After all, it takes considerable fuel just to keep a generator going, and cutting your power use in half will certainly not cut your expenses in half.

How Much and What Type of Power do you Need?

Systems come in all sizes and can do all kinds of things. But first, let's think about what kind of service we probably won't get from our renewable energy system. If there are any limitations at all on your level of service, then heating will be the task that is excluded.

Energy comes in many forms, from electricity to open flames. It can be divided according to its unruliness into high grade and lower grade energy. Electricity is considered a high grade energy source, because it is orderly and can do anything. An open flame is considered a lower grade energy source, because there are many things it cannot do, like charge up a laptop.

Heating requires only low grade energy. Any source, like the combustion of wood or propane, will do this task; whereas things like laptops and light bulbs use only high grade energy such as electricity. Why use a high grade source like electricity, carefully created, to do jobs that only require low grade energy, like heating?

When you take the attitude of service providing, it may be better to use some technology other than electricity for some jobs. While you can use high grade energy, like electricity, for all tasks, there just isn't any way to get a laptop to run on propane.

You may often find yourself, off-grid, using technologies other than electricity (such as wood or propane) for heating and refrigeration, unless you have a very large system. Heating, because it requires only low-grade energy, will always be the lowest priority for your renewable energy system.

Despite all this, assessing how much power you need to consume can be easy.

Assessing Your Consumption

The Consumption Self-rating Scale

In order to illustrate what you can do with various amounts of power, we use the problem of providing services to a remote household as an example. We will divide the different levels of service available into groups according to the Consumption Self-rating Scale. In order of increasing power, these groups are:

1: The basics
2: The essentials
3: Modern conveniences
4: Hot water heat
5: Space heating
6: Full-service household

How Much Power Do You Need?

1: The Basics	If you are a "one," you would be a weekend user, who would like to have a light per person, for a couple or three people, and a light for the room. This is a minimum lighting rule. A pickup camper or other similar small RV user might be a "one."
2: The Essentials	You might be a longer-term user, who might spend much more than weekends off-grid. As well as a light or two, you would like the use of some smaller appliances, such as a stereo, blender, or laptop computer. This level is frugal, but is acceptable for long term use.
3: Modern Conveniences	This level, for permanent residents, also requires tools and appliances such as refrigeration, freezing, and perhaps a well pump. This level includes all modern conveniences except those which require a lot of heat (e.g., a stove).
4: Hot Water Heat	This level provides modern conveniences and some heating of domestic hot water (enough for a large household).
5: Space Heating	This level includes all of the above, plus some space heating.
6: Full Service Household	This level also runs electric stoves, clothes dryers, and other high-draw appliances. (Generally, "more power than you know what to do with.")

Review the table, How Much Power Do you Need? and rate your power requirements from "the basics"... to "full-service household," which could also be described as "more power than you know what to do with."

Interpreting Your Score

1: If Your Rating is "The Basics"

If your rating is a "one," it may be that a photovoltaic system will fulfill your requirements. On the other hand, you may be attracted to microhydro for its other virtues. For example, microhydro resources tend to be quite generous. Almost any microhydro system will probably give more than a dozen or two kilowatt hours per month without much difficulty.

Maybe you are running an alarm system, or some other kind of instrumentation. Maybe you just want a light or two for a weekend cabin. While virtually any microhydro system will provide power at this level, which corresponds to just a few continuous watts, it may be that photovoltaics would be more suitable for providing small loads like this. A photovoltaic system is much easier to install than microhydro; a small but complete PV system with 80 peak watts of PV will cost about the same as the smallest microhydro system.

A complete microhydro system costs at least as much as a 120 peak watt PV panel. It is in fact more likely to cost a bit more than 300 watts of PV, and up, for most kinds of sites, just for the turbine.

With hydroelectricity, you must service intakes to keep them free of debris. Frost protection may also be necessary, and that can be a lot of work. Hydroelectric systems have, at the very least, one moving part. Some jobs, such as running a few instruments, just don't take enough power to make this kind of investment reasonable. Microhydro has certain fixed costs like the turbine and alternator, and they only come in certain sizes. You may have to pay for a unit with a 1,000 watt capacity even if you only generate 25 watts.

For a rating of "one," the 80 watt PV system mentioned above may provide all the power needed.

Photovoltaics is easy. There are no winterizing issues except for snow. It will work well nearly everywhere. Plus, of course, the panels are easy to install, relatively maintenance-free, and their only difficulty is that they are expensive. PV is also modular; you pay only for the capacity you get. If your power requirements are low, just a few watts, you only have to pay for a few watts of generating capacity. If your needs fall between March and September, so much the better.

> Virtually any microhydro system will provide power at level one, but it may not be the most appropriate system. If you need just a few kilowatt hours a month, say a dozen or two, you might better be served by photovoltaics.

> The frugal lighting rule: One light per person, and one light for the room.

By contrast, even the lowest cost microhydro system needs certain specific conditions of flow and head. Other systems for other kinds of sites require, in addition to a turbine and alternator unit, a diversion controller and load, a battery/inverter subsystem, and plumbing.

You may need structures that take the water out of the creek, remove debris, and then move the water to your turbine. These may be quite simple for the long pipelines characteristic of higher head systems, but higher flow rates may need significant structures. These fixed costs may be several thousand dollars at least, unless an existing pipeline and intake are used. Unless the specific head and flow characteristics are available for a low cost system, this fixed cost is considerably more.

While a certain PV system may have a given output in average months, there may be only a few dozen hours of usable sunlight in December. Performance is at its lowest in the winter, when the lighting loads are the highest and the solar resource is the lowest. Running in the winter is one of the virtues of microhydro that makes it worth some extra trouble. If you require power all year long, microhydro may be the best option, despite your low monthly requirements. And microhydro can be as simple as unreeling a couple of hundred yards of poly pipe and connecting a turbine, as in our case study "A Simple System," or lashing a low head turbine to a log at a convenient rapids, as in the case study "A Low Head AC System." Despite their casual nature, systems like this have a way of running for years.

Most PV systems use a backup of some kind, either a fossil fuel generator, or wind or water turbine. And while PV has no moving parts, all these backups do. A microhydro or wind turbine might only have one moving part, if you're lucky, but there are many, many moving parts in a fossil fuel generator.

Winterizing microhydro systems will produce a resource that operates when lighting and other demands are highest. And chances are that you will have more power as a result than with other systems.

Some places don't have freezing issues, but the water source may dry up in the summer. Microhydro is a logical choice to team up with PV in this situation. A hybrid system like this can mean that you provide all of your energy needs from renewables.

2: If Your Rating is "The Essentials"

If your rating is a "two," and you want "lights and music," even the smallest microhydro systems will give the kind of service you need. Look at the case study "A System With A Long Pipeline," and also at "A Low Head AC System."

What happens if you need more than just a tiny bit of power? What if you are staying in an off-grid place enough that providing good service becomes important?

An electrical system that provides "lights and music" is providing a good, minimal level of service. High-efficiency lighting and appliances means that many services that we get on the grid are now available off grid, if we make a few exceptions. Many people living off-grid will find this level of power meeting their needs very well. Heat, which is the exception, can be very satisfactorily provided with other technologies, such as propane or wood.

People report that an output of 35 to 70 kilowatt hours per month (that is, a system rated at 50 or 100 continuous watts) provides excellent "lights and music," the essential level of service. This level of power would deliver significant, if not unlimited, lighting, radio, stereo, radio telephone, kitchen appliances like food processors, TV, computer and VCR time to a remote household. Compared to the performance of even medium sized photovoltaic systems in the dark months, these are generous amounts of power.

There may not be enough power left over to power electrical refrigeration and freezing. And, of course, the amount of heating to be done is strictly limited to an occasional microwave or coffee maker. The choice of inverter will limit performance here more than any other factor.

3: If Your Rating is "Modern Conveniences"

After "lights and music," the next level of service that is commonly desired for the off grid household is electric refrigeration, freezing, water pumping and generally all modern conveniences that don't require much heat.

If your rating is a "three," and you want some modern conveniences, you can still get a powerful system from small resources. Efficient appliances can provide many modern conveniences to a household from as little as 75 to 125 kilowatt hours per month. A few hundred kilowatt hours per month can power even fairly ordinary appliances, with a big enough

A microhydro system that has an output of 75 continuous watts delivers 50 kilowatt hours per month, which rates as a "two" on the Consumption Self-rating Scale.

If you use efficient appliances, you can get a "three" level of service from a hundred or so kilowatt hours per month.

21

inverter. Look at the case studies "A Simple System" and "A System with a Long Transmission Line," as well as "A Small AC System."

To provide these services and start these appliances requires both kinds of power potential — both enough kilowatt hours per month on average (as in the Consumption Self-rating System), as well as sufficient inverter capacity (as in the Capacity Self-rating System). Where lights and music can be provided with a small, say, 500 watt inverter, adding large inexpensive appliances such as refrigeration and freezing will require more robust equipment.

Satisfying needs that are rated at one, two, and three can be accomplished with battery charging systems, as well as AC systems.

There are, however, excellent high efficiency electric refrigerators and freezers available that take far less power to run than ordinary appliances. For example, while the terribly ordinary refrigerator that came with our rented two-bedroom bungalow in Victoria, BC, uses 140 kilowatt hours per month, the Vestfrost refrigerator uses about 27 kilowatt hours per month. There are refrigerators, such as the Sunfrost, that are even larger, more efficient, and more expensive.

For example, an efficient refrigerator will start and run on a 1,000 watt inverter. In this situation, a 1.5-kilowatt inverter would provide many other modern conveniences as well as refrigeration and freezing.

While an energy efficient refrigerator may take only a 1,000 watt inverter to start, more ordinary appliances will require a 2.5-kilowatt inverter or larger to give the best service. Most washing machines can be used with these larger inverters as well. Another advantages of a larger inverter is that other loads with a high starting surge will be feasible as well. This will include many power tools. Many useful tools don't take that many kilowatt hours per month because they are only in use for a few hours, but can require a significant starting surge. (An "easy start" kit, which costs less than $25 (Cdn$40), can be added to motors to reduce their starting surge.)

A 4-kilowatt inverter will provide deluxe service for a household or light commercial applications. After the inverter capacity is available, more kilowatt hours per month of consumption are required to operate the appliances after they are started. A very ordinary refrigerator or freezer can use up to a couple of hundred continuous watts, once the various inefficiencies are figured in.

Remember that propane refrigeration is a reliable and effective alternative. It does, however, use quite a bit of propane, up to about a pound or more a day.

4: If Your Rating is "Hot Water Heating"

After filling needs for lighting, electronics, refrigeration and shop tools, the next task that your hydroelectric system may do is to heat your water.

If you are considering heating significant amounts of domestic hot water with your microhydro system, you will need to have an AC system, which means that you are dealing with a system that may use many times the water of a battery charging system. It can take a lot of kilowatt hours to provide a significant amount of domestic hot water. In Canada, the average power required to heat hot water is one third to one half of a very large bill. Put another way, the energy required for hot water heating can be many times the power an efficient household uses. It is often more effective to use solar hot water heating to capture low grade heat like domestic hot water, than to go to more trouble and expense to generate high grade electricity to make low grade energy. Examine the issue carefully.

After all, there just isn't that much heating capacity in a kilowatt hour, actually only 3413 BTUs. Basically, on-grid or off-grid, it will still take the same number of kilowatt hours per month to heat up a certain volume of water. With the exception of the heat pump (which is seldom used for domestic water heating), there is really no such thing as an energy efficient hot water heater, unless you mean solar. Demand hot water systems may have lower standby losses than the tank style ones do, but they still require that very large amounts of power be available while the water is running. This total can easily be more than all the rest of the power requirements of your energy efficient home put together. Solar water heating can be had for as little as $0.60 (Cdn$1) per peak watt.

Hot water tank recovery time is also important to think about. An ordinary immersion electric hot water heater has the capacity to use several kilowatts to heat up water quickly after it has been used. Without this peak capacity, actual hot water service performance may suffer. It may take overnight or longer to return a tank to usably warm temperatures after it is used.

Solar domestic hot water heating will provide a lot of heat units at reasonable cost. Winter hot water can be provided by using a special hot water heating coil in the firebox of a wood stove. Solar and wood heat complement each other well in the off-grid environment.

5: If Your Rating is "Space Heating"

As systems get larger, they can also supply significant amounts of space heating. Rating your needs at five will require a full sized AC system, such as the case studies on page 121 and 128. A thousand or two kilowatt hours per month is needed for this kind of service. At this level of power, most if not all appliances run without a question, and there are considerable seasonal surpluses.

6: If Your Rating is "Full-Service Household"

Level six, where you have "more power than you know what to do with," will be required in order to have enough capacity to run ordinary, high current appliances like ranges and electric dryers. A system that is this large will have surpluses most of the time, which could be sold. It takes a site like the case study "A Relatively High Output System" and an output of many thousands of kilowatt hours per month to get this kind of service.

Introducing the Capacity Self-rating Scale

Now that you know how much power you need, in terms of kilowatt hours per month, you also need to know how much power you will need at one time, your capacity, to help visualize the kind of system that would be right for you.

This is necessary to correctly size the inverter for DC battery charging systems, or to help decide how much capacity you will require from an AC system.

It's the same idea as the Consumption Self-rating Scale. Think about the largest load you will realistically need. Remember, the larger this load is, the more it costs. From the table on the next page, rate yourself from one to five.

The Capacity Self-rating Scale

ONE   Just the tiniest bit of power, to charge a laptop, for example
TWO   A few lights
THREE Full service, but little hot water
FOUR   A full-service household
FIVE   Enough to spare

If you just want a bit of AC power, rate yourself a "one" on the Capacity Self-rating Scale. Even a 50 watt inverter will charge a notebook computer, which uses 30 watts or so, as well as running an energy efficient task light. Other small inverters in this range can have outputs from 75 to 250 watts.

Rate yourself a "two" if you want to run a few efficient lights and other small loads. A 300–1,000 watt inverter is right for you. A sine wave inverter in this range will power many small appliances and sound systems. The "Long Pipe Line" case study had this level of AC power from its small inverter.

A 200-watt AC system could provide either level one or level two service; likely even level three.

If you need almost as much power as you can get from an extension cord, rate yourself as a "three." This means an inverter of about 1.5 kilowatt capacity is best. This will cost three or more times as much as the smaller inverters above, but you will be able to have efficient refrigeration and freezing if you have sufficient monthly capacity in kilowatt hours.

It takes an output of several hundred watts, up to a kilowatt and a half or so, of AC power to provide this kind of service. In the case study "A Small AC System," we ran an ordinary refrigerator on about 600 continuous watts.

A rating of "four" means that you would like to have a full-service household with regular appliances. A 2.5-kilowatt inverter is the minimum size that will start and run inefficient regular appliances. A sine wave 2.5-kilowatt inverter is an excellent all-around inverter, at about 50 percent extra in cost over the 1.5-kilowatt inverter that worked so well at the previous level. The case studies "A System with a Long Transmission Line" and "A Simple System" have capacities in this range.

The rule is the same
off-grid as it is on-grid:
It is usually easier
and less costly to
get better service by
conserving through
efficiency rather than by
generating more power.

If you have additional loads, such as domestic hot water heating and well pumps, which may require more than household quantities of power to start and run, rate yourself as a "five." The 4-kilowatt sine wave inverter is the answer. If you need an even larger configuration, stacking two inverters to double their output is a viable alternative.

Heaters come in all sizes and any heat you get from a battery charging system will likely come from the diversion load rather than the inverter, so heating requirements are really more about consumption than about capacity.

Off-grid and Needing Some Power

Being a significant distance from the grid and needing a bit of power is the best opportunity for microhydro. Off-grid, you could certainly use some lights and other services. In town, you may see by inspecting your power bill that you use maybe a thousand or more kilowatt hours per month. This does not, however, mean that you have to use the same amount of power to get the same service when you generate your own. For example, energy efficient lighting technologies use a quarter the power of more conventional light bulbs.

If you have a gravity water system that runs in the winter, you can have power, too. Credit: Scott Davis.

How Much is Enough?

The point is to pay attention to getting the service you require, from whatever source is appropriate, and not to worry too much about just how many watts are produced. In contrast to your expenses on-grid, where you pay for the amount used, off-grid you pay for capacity as well as for consumption. And so it pays to know when to quit.

People's habits vary quite a bit. People who use photovoltaics have the most efficient existence, whereas people with generators tend to use a lot of power to do the same task, even though they are paying more than a dollar or more per kilowatt hour for their power.

Do You Need Power in the Winter?

These methods still do not fully express the different kinds of service you get from different kinds of systems.

PV is expressed in average kilowatt hours per month. This means that between the equinoxes, that is to say from March 21 or thereabouts to September 21 or so, the system will give at least the rated amount of power. In the summer months in the middle, systems will perform significantly better.

However, on the other side of the equinoxes, between September 21 and March 21, a PV system performs less than average. Both the amount of sunshine available and the good weather that makes it possible to capture PV energy can be very scarce in the winter months.

In the darkest months, a PV system may only generate about 5–25% of its average power level.

If you have a summer cabin or RV, this "dark side" of the year is not important. The ease of installation of PV might recommend itself here. However, if your site is suited to microhydro, there may be many reasons to go to the trouble of installing a system. A microhydro system produces a steady supply of power all the time that the water runs.

Microhydro will deliver unmatched bang per buck when pairing an appropriate site with a suitable load. The steady output of a microhydro system means that battery storage requirements are minimal. If you really must live without batteries, AC power plants are available with outputs of 200 watts (that is to say, 144 kilowatt hours per month) and up.

If You Need Just a Bit of Power

Even the smallest microhydro systems, at about 18 kilowatt hours per month, will produce much more power than many photovoltaic systems. This may be important when providing lighting in the winter, when the solar asset is at its minimum and the requirements are at a maximum.

Whether power is produced from sun or water, 35–75 kilowatt hours per month, which is a typical small hydroelectric output, will give energy efficient lighting and electronic service, such as running a radiotelephone or stereo.

If You Live There Full-time

Many microhydro systems produce 75–350 kilowatt hours per month, which, with a proper inverter, will add large appliances such as refrigeration and freezing to a remote household. Our case studies on page 108 and 112 are in this range. Remember, efficient appliances will do more with less.

Magic Numbers

Happy customers report that a couple of hundred kilowatt hours a month, say from a battery charging system rated at 300 continuous watts, allows a household to use ordinary and inexpensive appliances and lights, given a large enough inverter.

Figure on half this amount to power high-quality, energy efficient (expensive) appliances and lighting.

After about 300 kilowatt hours per month, you don't get new and interesting kinds of service until power levels increase dramatically. We found that heating significant amounts of domestic hot water takes many hundreds of kilowatt hours per month.

Getting Even More Performance from Your System

As power outputs increase, a system will deliver adequate peak power as well as enough average power, and the battery and inverter subsystems are not required. This offers advantages, although generally an operator will have to install more and larger pipe, handle more water, and go farther to the intake to maintain a system with a larger output. AC systems make the same kind of power as you'd get from the grid. This sine wave power, as it is called, is available from PV systems, but you must use premium inverters to get this kind of service. Electronic load control has made governing AC easy and effective. Since hydroelectric systems put out the same amount of power all the time, an AC system must be large enough to provide not just the average output required, but the peak output required as well.

That said, there are wildly differing accounts of how big an AC system has to be to provide good service. I have heard professionals suggest that a system that produces 3,600 to 7,200 kilowatt hours per month (that's a 5 or 10 kilowatt continuous output) is required to power a household, and yet the smallest Powerpal turbine, which is designed for electrification in rural areas of developing countries, puts out 144 kilowatt hours per month. Our case study "A Low Head AC system" produces 720 kilowatt hours per month, to light a village of 23 households and a school.

So, as you see, there is some disagreement about this topic. The case study "A Small AC System" explains the strengths and weaknesses of this kind of system.

When systems get larger, and a few hundred extra kilowatt hours per month are available, they will make a significant contribution to hot water heating.

And again, once you have provided lights and music, large appliances, and some domestic hot water, each additional kilowatt will probably be used for space heat. Each continuous kilowatt heats a little better than a cord of firewood; in fact, for a six month heating season, 10 kilowatts of electricity is equivalent in heating capacity to 12 cords of firewood.

So there you go. Now that you know what to expect, how do you find it?

3

Assessing Your Site

IN THIS CHAPTER, we'll discuss how to measure the important characteristics of your site, and how to calculate its power potential.

Water flow and water pressure together make power. Assessing the potential of a site means assessing the power potential of a pipe, whether it is one that is already in place, or one that you are considering using.

On the other hand, and this is their great advantage, some low head systems may not use much pipe, except perhaps for the draft tube, a few feet long, which may be included in the machine. Water may be brought to the unit with a flume, or there may be an existing structure to provide the small drop necessary.

But most microhydro systems use pipes. To calculate the power you are going to have you must find out:

- How much water flow is available.
- How much static pressure or static head is available.
- The length, diameter and material characteristics of the pipeline required.

When these things are known, then you can calculate:

- **Choosing the water flow rate in the pipeline**. As water begins to flow, then some of the pressure is used up as friction. The pressure available for generating goes down as the flow increases. Eventually, diminishing returns set in, and you should quit running more water through the pipe. You will have reached a point where more flow will not do you any good as far as generating power is concerned. This point occurs when about a quarter to a third of the potential pressure in the pipe is lost to friction. To get more power from the water they use, many systems use pipe that has frictional losses in the 10 to 15 percent range.

- **Net head at this flow rate**. The net head is the pressure that your pipeline has when the water is running. You might be trying to get the most out of your pipe, or you might be optimizing the potential available from a given amount of water, or you might need to reduce the load on your intake while still making plenty of power, or you might want to run lots of water through the pipe for frost protection. Let's call the flow rate that meets your needs the "optimum" flow rate.

When you know the net head available, and the flow rate for your pipe, the power can be easily calculated.

This book has the charts and formulas you need to choose the size of pipe you need, estimate how much water you will use, and get a good idea of your potential output.

Step 1: Measuring Stream Flow

The first thing you need to know is the total amount of water available.

If you use a spring or a very small creek, it may be that the amount of water available is a factor that limits your power output.

One secret of water flow measurement is that it is seldom indeed that a whole creek is taken up with a microhydro project. None of our case studies use all the water available to them. Piping is expensive and it is much more likely that economic considerations will make pipe diameter and length the factors that limit the amount of water used in a system.

> Seldom indeed is a whole creek taken up with a microhydro project.

Microhydro systems in this book use between 2 and 1,000 gallons per minute. We use US gallons for calculations, because most references use the US gallon.

Estimating Flow Rate

One perfectly good way to measure flow is to compare your flow to the four pictures of a known flow seen on the facing page. Note the water levels in each of the photos, in order to get an idea of flow.

Over 500
US Gallons

Under 400
US Gallons

Top Left: Over 500
US gallons per minute.
Credit: Bonnie Mae
Newsmall.

Top Right: Under 400
US gallons per minute.
Credit: Bonnie Mae
Newsmall.

About 100
US Gallons

About 20
US Gallons

Bottom left: About 100 US gallons per minute. Credit: Scott Davis.

Bottom Right: About 20 US gallons per minute come from this
machine. Credit: Ann Cavanagh.

The Container Method

low in gpm equals the size of the container in gallons, divided by the time it takes to fill in seconds, times sixty.

Another way to measure water volumes is the "container method," for use where flow rates range up to a few hundred gallons per minute.

To measure the how much water there is in a creek, find a spot where the water enters a culvert, and time how long it takes to fill up a container of a known size. Flow in gpm equals the size of the container in gallons, divided by the time it takes to fill in seconds, times 60.

For example, if your 5-gallon bucket fills up in 10 seconds, you have 30 gpm.

Some sites may fill a bucket in a flash. This means that there are hundreds of gallons per minute available. If more accuracy is needed, use a larger container, such as a 45-gallon drum.

The container method will be appropriate for most high head systems. They use up to about 200 gallons per minute, which fills up a 5-gallon bucket in about a second and a half. This photo shows a flow much higher than that.

Sometimes your bucket will fill in a flash.
Credit: Scott Davis.

The Weir Method

For systems that will use more than a couple of hundred gallons per minute, using the weir method will give good results. A weir is a notch of a given size and profile over which water flows. By measuring the depth of water, and knowing the size and shape of the notch, quite accurate measures of water volume are possible.

For example, the LH-1000 requires a little over three inches of water flowing over a weir five feet wide, or 1,000 gallons per minute, for its maximum output of 1,000 watts. It will produce useful amounts of

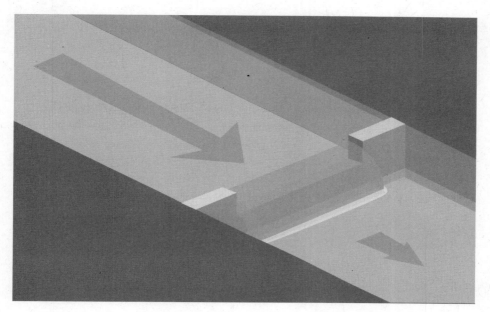

Measure the depth of water flowing over a weir. Credit: Corri Loschuck.

Table 3.1: Weir Table

inches of head	1 foot wide	3 feet wide	5 feet wide	gpm/foot over 5 feet wide
1	35	107	179	36
1.5	64	197	329	66
2	98	302	506	102
2.5	136	421	705	142
3	178	552	926	187
4	269	845	1420	288
5	369	1174	1978	402
6	476	1534	2592	529
7	–	1922	3255	667

power down to 2 feet of head and 500 gallons per minute. The smallest Powerpal uses three inches of water flowing over a weir 3 feet wide, or 550 gallons per minute.

Step 2: Measuring Pressure

Next, you have to measure the total amount of water pressure that will be available to you. The pressure is highest in a pipe when no water is moving; this is the "static pressure."

You discover how much pressure you will have by measuring it directly or by surveying the distance that the water drops in feet (the head), and calculating the pressure.

Using a Pressure Gauge

You may already have a pipe in place that could become the penstock.

Pressure in psi equals feet of head times 2.31.

The easiest way to measure head is to measure the pressure in an existing pipe and then convert pounds of pressure to feet of head.

Many places have pipelines in place already. Gravity domestic water is often supplied to dwellings with pipe large enough and pressures and available flows high enough to make good penstocks.

If you have a pipe in place, install a pressure gauge. (All systems should have a pressure gauge anyway, to help with troubleshooting.) The pressure in pounds per square inch directly reflects the vertical distance in feet from the gauge to the surface of the water in the intake.

Since only about a third of the static pressure in a pipe, at most, is used for microhydro, lots of pressure is usually left over in the pipe with which it can continue doing its original tasks.

To assess the pressure in an existing pipeline, first turn off all running water. Use a gauge to measure the pressure. A temporary installation can be done by attaching a gauge to a garden hose faucet (called a "hose bib") if there is one that is appropriate. This will yield the static pressure, and from it you can calculate the static head.

Measure pressure, and thus head, with a gauge.
Credit: Scott Davis.

Converting Pressure into Head

When you measure pressure with a gauge, it is easily converted into head by multiplying the pressure in pounds per square inch by 2.31. For example, 100 psi = 100 × 2.31 = 231 feet.

And, of course, the opposite is true: divide the head in feet by 2.31 to find the pressure. If you have 231 feet of head, you have pressure of 100 pounds.

Measuring Maximum Flow Rate

The maximum flow rate can be measured directly. After you have taken the static head, the net head will be about two thirds to three quarters of the static head. Open the valve until the pressure drops to about this figure, and measure the volume of water flow that creates this net head.

Or, again, if you have more pipe than water, turn on the valve until you use all the water you can. Note the pressure. This will be your net pressure.

This method of dynamic testing is particularly valuable with very long pipelines, to confirm that the actual hydraulic characteristics of the pipe are the same as you thought they were. Pipes that have been buried for a long time may have forgotten features, such as repairs or corrosion, that would affect the performance of the pipe. Even if you can't remember precisely all the details of pipe characteristics, this method will reveal the actual pipe performance.

Surveying

If you don't have an existing pipe, survey the potential site. Surveying physically measures the vertical drop. Then, measure the length of a proposed pipeline with a tape, hipchain or pedometer. Measure the drop in feet. When you have measured the drop between the top and the bottom of your potential penstock, you have the static head. This drop in feet is what the pressure gauge will measure in pounds per square inch when the pipe is full and there is no water moving.

Measuring the head is like measuring length along the ground, except that you are measuring vertical distance. If you are starting with a stream and a hillside, the measurement needs to be accurate enough that proper

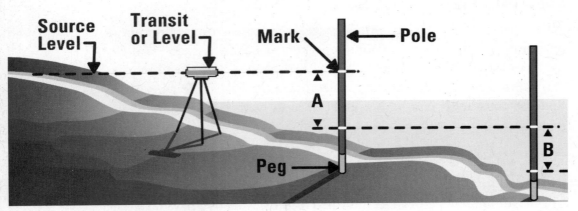

How to survey. Credit: Corri Loschuck.

pressure ratings for the pipe are assured. The accuracy of the power estimate and jet sizing also requires measurement of head within five percent or so. Low head systems require careful measurement to make sure that they will operate as designed.

I have successfully measured high head systems, over 300 feet, with the altimeter that comes with my watch. This rough estimate was adequate to mark the place where pipes of increasing pressure rating had to go.

This method was accurate within ten percent or so. This was adequate because the nozzle was variable and could adjust to a wide range of sizes.

The more accurate way to measure vertical distances, and one that should be used if you don't have the luxury of a variable nozzle, is to use a rod and a level to measure the elevation as you work your way up a hillside. This is an ancient way of measuring that still works fine today. This method will give results that are good enough that you can calculate with accuracy the size of nozzle your turbine requires.

To measure using this method, you need a measuring rod, level and staff. Begin by first making a measuring rod that is a convenient length, say eight feet long. Mark the feet off in bright colours, with tape or paint.

A staff supports the level. It needs to be a convenient length to hold a level on the top and for you to sight over, say four or five feet. A hand level is a surveying instrument that can be operated by one person. You can use a hand level or even a carpenter's level, although a carpenter's level will need a third person, since one person cannot sight down the level the long way and see the bubble from the side at the same time.

Begin at the starting point by having one person hold the measuring rod vertically. The person (or people) with the level then goes up the hill until the level, resting on the staff, is at the same height as the top of the measuring rod. With an 8-foot measuring rod, the top of the rod, and the level, are now 8 feet above the starting point.

Then, the person (or people) with the level swings the level around and sights and marks a new point on the hill that is level with the top of the measuring rod and with the level.

The person with the measuring rod moves to this point. Then they set the bottom of the measuring rod just at this mark. The bottom of the measuring rod is

Surveying. Credit: Ann Cavanagh.

now at the same height above the starting point as the top used to be. The top of the measuring rod is now 16 feet above the starting point. Work your way up the hillside, carefully recording a running total, to measure the total drop. (It may be that in places brush or other obstructions make it difficult to use a full 8-foot rod. That is why the feet are marked.)

Using a Water Level

You can measure head, especially for low head systems, very accurately with a water level. This is a fancy name for a piece of pipe. Put one end in the water and let the hose fill. You may have to encourage it a little. When water is flowing out the lower end, pick it up higher and higher. Eventually, water will just stop flowing from the pipe. Measure this distance from the ground and that is the distance that the water has fallen.

Repeat as necessary and add up the distances if required.

Go Figure!

Here are some of the facts about pipelines in the case studies which appear later in the book. Calculate their net heads and pressures:

1. From the case study "System with a Long Pipeline" – 65 psi of static pressure

2. From the case study "A Simple System" – 45 psi of static pressure

3. From the case study "System with a Long Transmission Line" – 75 feet of static head

4. From the case study "A Small AC System" – 112 feet of static head

5. From the case study "A Relatively High Output System" – 315 feet of static head

1. $65 \times 2/3 = 43$
$65 \times 3/4 = 49$ (or so – this isn't rocket science!)
The net pressure of this system should be between 43 and 49 psi.
$43 \times 2.31 =$ about 100
$49 \times 2.31 =$ about 113
The net head on this system will be in the range of 100 to 113 feet.

2. $45 \times 2/3 = 30$
$45 \times 3/4 = 34$
The net pressure will be 30 to 34 psi.
$30 \times 2.31 = 70$
$34 \times 2.31 = 79$
The net head will be 70 to 79 feet.

Step 3: Creating a Stream Profile

A stream profile combines information you have just gathered about the drop available in a creek with a measurement of the length required. You can measure length by various means such as tape measure, hip chain, or GPS. Just keep careful records of how far it is.

The completed stream profile will sound like "The first 100 feet drops 20 feet. The second 100 feet are not as steep, and drops 16 feet" and so on.

By surveying a stretch of creek, you may find places where it is steeper than others, as well as identifying places where it would be easiest to work.

3. $75 \times 2/3 = 50$

$75 \times 3/4 = 52.5$

The net head will be 50 to 53 feet.

$50 \div 2.31 = 22$

$53 \div 2.31 = 23$

The net pressure will be 22 or 23 pounds. Remember, your pressure gauge is only accurate to the nearest pound or two.

4. $112 \times 2/3 = 75$

$112 \times 3/4 = 84$

The net head should be between 75 and 84 feet.

$75 \div 2.31 = 32$

$84 \div 2.31 = 36$

The net pressure should be between 32 and 36 psi.

5. $315 \times 2/3 = 210$

$315 \times 3/4 = 236$

The net head could be between 210 and 236 feet.

$210 \div 2.31 = 91$

$236 \div 2.31 = 102$ psi

The theoretical net pressure should be 91 to 102 psi. In this case, the penstock was so short compared to the diameter and the amount of water available that the jet on the turbine was not big enough to reduce pressure very much at all.

Remember, most of the maintenance is at the intake and so the intake should have good access.

For very small flows of water, there may only be one turbine that is the best. But as soon as flows get above a few gallons per minute, there may be a choice of possible turbine/alternators to use. In order to assess which would be best, a complete knowledge of the watercourse will make all the different alternatives clear.

As a rule of thumb, electricity is easier to move than water. It can go uphill, too, so that a turbine could be located below where the power is

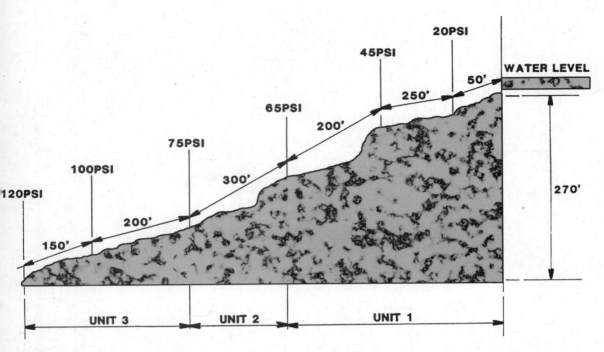

A stream profile measures the vertical drop and the distance over the ground that your pipeline might take. Credit: Corri Loschuck.

used, if that turns out to be convenient. As waterflows get to the 500 gpm level and above, many more alternatives may be possible. Since the penstock can be the most difficult part of any system to install, it may well be worth using a low head system and short penstock, located in a part of the creek that is a little steeper than the rest, and moving the electricity to the point of use.

Although this plan would use more water, penstock construction effort and cost may be significantly reduced.

Calculating Potential Output

After you have made your measurements and created a stream profile, you are ready to go to the next stage by calculating how different pipes would do as penstocks.

Step 4:
Calculating Allowable Frictional Loss Rate

As water begins to flow, friction reduces the pressure available. As more water flows, there is more friction and the pressure continues to drop. The length of pipe determines the total amount of friction from a given flow. Thus, a certain flow produces double the pressure loss in a 1,000 foot penstock than in a 500 foot penstock.

As it happens, the maximum amount of power comes out of a penstock when a flow of water causes about a quarter to a third of the static head to be lost. If only a certain flow is available, a larger pipe will have lower friction. A higher output from the same length of pipe will result from increasing the penstock size. However, price goes up accordingly. Determine your costs carefully and compare them to the benefits gained. Is it worth it? Think about it.

In many applications, there is a lot more water available in the stream than an affordable pipe can handle. In that case, jet sizes are chosen in order to deliver the largest flow of water to give an acceptable pressure loss.

When the amount of water available is less than this amount, the frictional loss rate will be a product of volume and length of pipeline. Since pressure and head are equivalent, pressure loss is measured in feet of head lost per hundred feet of pipe. Thus, a long pipeline delivers less water to your turbine than a short pipeline at the same pressure. Generally, a pipeline has few fittings, elbows, reducers and the like, but their influence on frictional losses is equivalent to adding more pipe.

To calculate the optimum frictional loss rate with lots of water available, you must first know how much head or pressure can be lost to friction. This gives you the maximum amount of pressure you can lose.

Relate this pressure loss to the length of the pipeline by dividing the total frictional loss along the whole length of the pipeline.

Express the length of the pipeline in hundreds of feet to get an optimal frictional loss rate expressed in feet of head loss per hundred feet of pipeline. Find this figure on the column of the appropriate frictional loss table on either page 46 (PVC) or 47 (polyethylene).

Go Figure!

Let's say you have measured a static head of 60 feet, in a run of 600 feet of 2" poly pipe. You need to choose a flow rate. A 25% loss to friction produces lots of power, using not too much water; the maximum you should use would produce a 33% frictional loss.

25% of 60 = 15 feet of head loss
33% of 60 = 20 feet of head loss

So then, the ideal head loss rate is between:
15 ÷ 6 (hundred feet of pipe) = 2.5 feet per hundred
20 ÷ 6 (hundred feet of pipe) = 3.3 feet per hundred

So the maximum head loss rate is between 3.3 feet per hundred and 2.5 feet per hundred. A lower rate of head loss will produce more power from a given amount of water. Always check and see if using the next size larger pipe is cost effective.

Step 5: Finding a Flow Rate

If you have lots of water available, here's how to find the proper flow rate for your particular pipe:

- Once you know the optimum head loss rate, go to the chart that describes your pipe (PVC or poly).
- Go down the column that covers your pipe size until you find the range of values that covers your optimum frictional loss rate.
- Your optimum flow rate will be in the column on the left, "Flow in US gpm."

It might have to be a range of values, but you will still get good results by interpolating.

If you are limited in how much water you have available, that's your optimum flow rate. You need to calculate the net head that this flow rate will have in your particular pipeline. This is how to do it:

Go Figure!

To continue our example from above, go find 3.33 per hundred in the "2 inch" column on the polyethylene chart. It falls between the values for flows of 40 and 45 gallons per minute.

Your optimum flow rate is between 40 and 45 gallons per minute.

If you were limited to a flow rate of 20 gallons per minute, what would your net head be?

Remember, you have 600 feet of 2" poly pipe at 60 feet of static head.

Look at **Table 3.3: Head Losses from Friction in Poly Pipe** (on page 47). Go down the 2" column until you come to the 20 US gpm row, and you will see that you will lose about 1 foot of head per hundred feet of pipe, using 20 US gpm in 2" pipe.

1 (foot of head loss per hundred feet of pipe) × 6 (hundred feet of 2" poly pipe) = 6 feet of head loss.

60 feet (static head) − 6 (feet of head loss) = 54 feet of net head.

If you had 600 feet of 2" poly pipe and 60 feet of static head, you would have 54 feet of net head with a flow of 20 US gallons per minute.

- Go to the appropriate chart and find the column that describes your pipe size.
- Find the row that expresses your given flow of water in the first column.
- Where the pipe size column meets the flow rate row, you will find the loss rate per hundred feet of pipe.
- Multiply this loss rate by the number of hundreds of feet in your pipeline.
- Subtract these losses from the static head to give the net head with your pipeline and flow.

Table 3.2: Head Loss from Friction in Class 160 PVC Pipe
(in feet of head per 100 feet) C = 130

flow in US gpm	pipe size in inches									
	1"	1.25"	1.5"	2"	3"	4"	5"	6"	8"	10"
3	1.0	.3	.1							
4	1.7	.6	.2	.1						
5	2.5	.9	.4	.1						
10	9.2	3.1	1.3	.3						
15	19.5	6.6	2.7	.7	.1					
20	33.1	11.2	4.6	1.1	.2					
30	70.1	23.7	9.7	2.4	.3	.1				
40		40.3	16.6	4.1	.6	.1				
50		60.9	25.1	6.2	.9	.2	.1			
60		85.3	35.1	8.6	1.2	.3	.1			
70			46.7	11.5	1.6	.4	.1	.1		
80			59.8	14.7	2.0	.5	.2	.1		
90			74.3	18.3	2.5	.6	.2	.1		
100			90.3	22.2	3.1	.8	.3	.1		
150				47.1	6.5	1.6	.5	.2	.1	
200				80.2	11.1	2.7	.9	.4	.1	
250					16.8	4.1	1.4	.6	.1	
300					23.6	5.8	2.0	.8	.2	.1
400					40.1	9.9	3.3	1.4	.3	.1
500					60.7	14.9	5.0	2.1	.5	.2
600					85.0	20.9	7.1	2.9	.7	.2

If you have lots of water available, the net head will likely be three quarters or as low as two thirds of the static head. With limited water flow, the net head is the static head minus the losses incurred by the volume of water available.

With these charts, you can easily find the optimum flow rate for various pipe sizes and types that are commonly used in microhydro development. You can also use the chart to assess the effects of using flows and different diameters of pipe. Often, it may be that using a larger pipe size is an easy and cost effective way to increase the output of a system. The lower friction of larger pipe will allow a higher optimum flow rate, or it will allow a given amount of water to flow with less loss (greater head).

Table 3.3: Head Loss from Friction in Poly Pipe
(in feet of head per 100 feet) C = 140

flow in US gpm	pipe size in inches				
	1"	1.25"	1.5"	2"	3"
3	.9	.3	.1		
4	1.5	.5	.2	.1	
5	2.2	.7	.3	.1	
10	8.0	2.7	1.1	.3	
15	17.6	5.7	2.4	.6	.1
20	28.9	9.7	4.0	1.0	.2
30	61.2	20.6	8.5	2.1	.3
40		35.1	14.5	3.6	.5
50		53.1	30.6	5.4	.7
60		74.4	40.7	7.5	1.0
70		98.9	52.1	10.0	1.4
80			64.8	12.8	1.8
90			78.7	16.0	2.22
100				19.4	2.7
150				41.1	5.7
200				69.9	9.7
250					14.7
300					20.6
400					35.0
500					52.9
600					74.1

Step Six: Calculating Power Output

To calculate power output, first calculate net head. If you have lots of water available, an optimum net head is two thirds of the static head. With limited water flow, the net head is the static head minus the losses incurred by the volume of water available.

Then calculate power output from net head and flow rate. Once you have the net head and flow rate, then assessing the potential output of a microhydro system, in continuous watts, follows this formula:

$$\text{Continuous watts} = \text{flow (gpm)} \times \text{net head (feet)} \div 10$$

This gives your output in continuous watts, using modern turbines and alternators. To calculate your expected output in kilowatt hours per month, multiply the flow rate in gallons per minute by your net head in feet, and divide by 13.8. Efficiencies may be a bit higher with larger AC systems.

To use this formula with other products that may not be as efficient, cut the expected output according to the following rules:

> To calculate your expected output in kilowatt hours per month, multiply the optimum flow in gallons per minute by your net head in feet and divide by 13.8.

- If you are using older turbines, cut the expected output by 25 to 50 percent.
- If you are using an efficient automotive alternator such as the Ford 70A, reduce output by about 40 percent. This means that power = flow × head ÷ 14. However if you are using an inefficient alternator such as a Delco, reduce your output by 60 percent. If you have an AC system, especially a larger one, increase your expected output by 10 percent, or even a little more.

Choosing Nozzles

Use Table 3.4 opposite to choose your nozzle size. Find your net pressure in the pressure column, and follow its row over to your desired flow rate. The nozzle diameter required is at the top of the column containing your flow rate.

Nozzle diameters are limited by the size of the runner. More water can be put through a given wheel by using more than one nozzle.

Nozzles do not have to be the same size. By turning nozzles of different sizes on and off, a wide variety of stream flows can be accommodated. This means that systems do not need to be limited to the lowest flow rate, but can have a nozzle that is small enough to accommodate low water times, and other nozzles which can be turned on when more water is available. The case study "Upgrading a Small AC System" has three nozzles, each one inch in diameter. Using two nozzles reduced the water flow significantly without making too much difference in performance. On the other hand, when the water was available, the third nozzle gave somewhat more power.

Table 3.4: Choosing Nozzles

nozzle size in inches

net pressure psi	net head feet	.125"	.1875"	.25"	.375"	.4375"	.5"	.5625"	.75"	.875"	1"	1.25"
2	5	–	–	–	6.18	8.40	11.0	14.0	24.7	33.0	43.7	–
4	10	–	–	6.05	8.75	11.6	15.8	19.5	35.0	47.6	61.6	–
6.5	15	–	–	7.40	10.7	14.6	10.6	24.0	42.8	58.2	75.7	–
10	23.1	–	3.32	5.91	13.3	18.1	23.6	30.2	53.1	72.5	94.5	148
20	46.2	2.09	4.69	8.35	18.8	25.6	33.4	42.4	75.1	102	134	209
25	57.7	2.34	5.25	9.34	21.0	28.7	37.3	47.3	84.0	115	149	234
30	69.3	2.56	5.75	10.2	23.0	31.4	40.9	51.9	92.0	126	164	256
35	80.8	2.77	6.21	11.1	24.8	33.8	44.2	56.1	99.5	136	177	277
40	92.4	2.96	6.64	11.8	26.6	36.2	47.3	59.9	106	145	189	296
50	115.5	3.30	7.41	13.2	29.7	40.5	52.8	67.0	119	162	211	330
60	138.6	3.62	8.12	14.5	32.5	44.3	57.8	73.3	130	177	231	293
75	170.2	4.05	9.08	16.2	36.4	51.2	64.7	84.4	146	205	259	404
100	231	4.67	10.5	18.7	42.1	57.3	74.7	95.0	168	229	299	467
135	311	5.43	12.2	21.7	48.9	65.1	86.7	107	195	261	347	543

jet flow rate in US gpm

Go Figure!

Try these problems and see how some real life situations have looked on paper, and consider various options. Find the expected power output in each of these situations:

From the case study "System With A Long Pipeline:"
What are the effects of using larger pipe on power outputs?

A. 1.25" poly pipe, 2,000', 65 psi
- First, convert pressure to head: 65 psi × 2.31 = 150 feet of head.
- Then, decide how much head you can afford to lose. A quarter of 150 is 38; a third of 150 is 50. You can lose from 38 to 50 feet of head in all.
- This means that the net head will be – theoretically – between 100 and 112 feet.
- Look at **Table 3.3** on page 47, under the 1.25" column. 7.5 US gpm gives a loss of about 1.7 feet per hundred. Over 2,000 feet, your losses will be 1.7 × 20 = 34 feet. This puts your net head at about 115 feet.
- 7.5 US gpm × 115 feet net head ÷ 14 = about 62 watts. So, getting 70 watts is a little better than expected. Maybe the pipeline is a little shorter than 2,000'.
- Try different jets about this size to see which works best.

B. 2" poly pipe, 2,000', 65 psi
- As above, the allowable head loss rate is between 1.9 and 2.5. This time, however, look under the 2" column in **Table 3.3** on page 47.
- A flow of 30 or more US gpm will give a head loss of about 2.1 feet per hundred feet of pipe, and so the head loss will be 2.1 × 20, or 42 feet. With a static head of 150 feet, this means that the net head will be 150-42, or 108 feet. 30 US gpm × 108 feet of net head ÷ 14 = 230 feet.
- As above, the net head is 100 to 120 feet.
- 30 US gpm × 10 feet ÷ 10 = 300 watts (or more; this is a conservative estimate)
- A really significant power increase would come from using 2" pipe rather than the 1.25".

From the case study "System With A Long Transmission Line:"
Does all the capacity of the pipe need to be developed?

20 gpm, 3" PVC pipe, 500', 60 psi
- First, look up the head losses due to friction in **Table 3.2: Head Loss from Friction in Class 160 PVC Pipe**, on page 46. In the row 20 US gpm, under the 3" column, you will see that about 0.2 feet of head is lost from 20 US gpm flowing in a 3" pipe.
- Since this pipe is 500 feet long, the head loss will be 5 (hundred feet of pipe) x 0.2 (feet of head loss per 100 feet of pipe) = 1 foot of head loss.
- Static head is 60 psi × 2.31 = 139 feet. With 1 foot of head loss, the net head is about 138 feet.
- The theoretical output is 20 US gpm × 138 feet of net head ÷ 10 = about 275 watts
- This particular setup has excellent hydraulic efficiency. A high net head compared to static head means that more of the energy in the pipe is used to generate power, and less is lost to friction.
- On the other hand, this site has significant electrical efficiency issues, stemming from the induction alternator and homebrew transformer system. But it works!
- In this case, hydraulic efficiency was quite inexpensive (the price difference between 3" and 2" pipe was just a couple hundred dollars).

From the case study "A Simple System:"
What is the potential output?

60 gallons per minute, 2" poly pipe, 600', 60 psi
- Calculate the static head: 60 psi × 2.31 = 139
- Look up the frictional losses of 60 gpm in 2" poly, which are about 7.5 feet per hundred.
- Calculate the total loss to friction: 7.5 (feet per hundred of frictional losses) × 6 (hundred feet of pipe) = 45 feet of head loss.
- Calculate net head: 139 (feet of head) − 45 (feet of head loss) = 94 feet of net head.
- Calculate the power output: 60 US gpm × 94 (feet of head) ÷ 10 = about 564 watts.
- Lots of potential there! In this case, a couple of hundred watts provided good service, and the full potential of the pipe did not need to be developed.

The following examples are from the case study "A Small AC System."
With a 5" PVC pipe, 500' long, with 112' of static head, how much power could be generated with the following flows and pressures?

A. Buck Creek 'B': 113 US gpm at 45 psi
- Calculate net head in feet: 45 psi × 2.31 = 104 feet of net head
- Theoretical output: 113 US gpm × 104 feet of head = 1,175 watts

B. Buck Creek 'C': 310 US gpm at 40 psi
- Calculate the net head in feet: 40 psi × 2.31 = 92 feet of net head
- Theoretical output: 310 US gpm × 92 feet of head ÷ 10 = 2,850 watts

C. Buck Creek 'D': 378 US gpm at 40 psi
- Calculate the net head in feet: 40 psi × 2.31 = 92 feet of net head
- Theoretical output: 378 US gpm × 92 feet of head ÷ 10 = 3,475 watts

D. Buck Creek 'E': 528 US gpm at 34 psi
- Calculate the net head in feet: 34 psi × 2.31 = 79 feet of net head
- Theoretical output: 528 US gpm × 79 feet of head ÷ 10 = 4,171

Some hydraulic efficiency was lost as flows through the turbine exceeded 200 US gpm. However, as with other sites, there was lots of power despite inefficiencies of one kind or another.

From the case study "A Relatively High Output System."
With a 4" PVC pipe, 900', with 275' of net head, how much power could be generated with the following flow and pressure?

- 350 US gpm at 135 psi, with 275 feet of net head
- 350 US gpm × 275 feet of head ÷ 10 = 9,625 watts
- Add 10% to this output figure, since this was a more efficient AC system.

4

The Appropriate System for You

T HIS CHAPTER WILL TAKE YOU FROM YOUR ASSESSMENTS OF YOUR NEEDS
and the stream profile and site assessment you have just done, to the
system that will be appropriate to meet your needs with the resources you
have.

Battery Charging Systems

If you have rated yourself at one, two or three on the Consumption Self-
rating Scale (see page 18), this chapter will explain how a battery charging
system could meet your needs.

Batteries in Microhydro

If, unlike other renewable technologies, microhydro is not intermittent,
then why use batteries? In North America, most microhydro systems store
energy in batteries and use inverters to make AC. The advantage of this
system is that it accumulates power and delivers it when needed. This
makes many small sites — which would otherwise not be very useful due
to low power outputs — into very good systems indeed. Rather than
providing peak power, the turbine only has to deliver the average power
required by the system.

Here's the question:
If, unlike other
renewable technologies,
microhydro is not
intermittent, then
why use batteries?

Charging Batteries

A major advantage of battery charging is that one turbine can be used for
a wide range of high, medium and even fairly low heads, down to about six

feet. There are also turbines that are designed specifically for low head battery charging systems.

Electrical use varies throughout the day. Although sites differ widely in their requirements, a major use of electricity in an off-grid situation (40% or more) might be lighting, which may only be on part of the time. Late at night, and all day, the turbine can be charging up batteries for their period of heavy use in the evening.

Microhydro turbines usually charge at the same rate all the time. This contrasts with photovoltaics, where the sun goes behind clouds occasionally, goes down at night, and goes south for the winter. Wind can blow by day and night, but then it can be still for many days, even at a good site, making it another intermittent resource. As a result, battery requirements for a microhydro system are much lower when compared with other technologies. Where a PV or wind-powered system might have a battery subsystem that will store enough energy for several days without usable sun or wind, the microhydro system only needs to store energy for a few hours. This means that the proportion of the system that is batteries is much smaller in a microhydro system. A small microhydro system might only use a couple of golf-cart batteries. While this kind of battery offers nothing new in terms of technology, it is readily available, relatively inexpensive, and is a known factor. These batteries are the key to making many thousands of otherwise unusable water sources into potential microhydro sites.

Also, as a rule, with a microhydro system, the batteries are discharged a lot less fully than in other kinds of systems, and are promptly recharged. This means that they will last longer. A microhydro system generally has its batteries recharged on a daily basis. Thus the amount of storage used is often less than a day's worth of power. Many times, the battery pack is sized to provide sufficient current for the inverter, rather than to provide some amount of storage capacity.

Batteries are not strictly required, even in small microhydro systems. If reduced performance from a small system is acceptable, it is possible that batteries can be eliminated if desired. Larger systems do not use batteries.

Microhydro tends to produce more power than is needed, and many battery problems stem from overcharging. Without proper controls, even a small current working day and night will boil batteries dry.

Batteries store up power for use when required. Usually, six-volt deep cycle golf cart batteries are used. Two are required for a small system, four or more for a larger one, depending upon the size of inverter that is chosen. The number of batteries you have is determined mainly by the size of the inverter that is to be run. A day or two of battery storage is usually plenty.

Charge Control

All battery charging systems require charge controlling. Eventually, the batteries get fully charged up. In a small, hard working system, it may not happen every day, but eventually it will. If the batteries continue to be charged, the energy that was being stored will be used to "boil" the electrolyte and eventually the batteries will dry out and be damaged.

Charge control prevents this battery damage from happening.

This is achieved by a device called a diversion controller which senses battery voltage. When the voltage is too high, indicating that no more charging is required, the diversion controller switches on a dump load.

The result is that, even if there are no other loads on at the time, the battery is not being overcharged.

Most battery charging systems operate at pretty low power levels. The diversion load is often quite small, but can still be used to keep the powerhouse or batteries from freezing or some other task. In battery charging systems it is not likely that enough energy will be unused to heat a significant quantity of domestic hot water, but other smaller loads, such as keeping a root cellar from freezing, may be practical.

Producing AC from Batteries with an Inverter

AC is a more useful form of electricity than DC, if only because it is more common. The drawback is that AC cannot be stored like DC.

The inverter makes it possible to use common AC appliances, and generally provides ordinary electricity in the same way we normally use it. Inverters don't take much to run and so they are fairly efficient for those times when no power is being used. They will provide peak power outputs many times their continuous rating for short periods of time to provide the starting surge that motors and other applications need.

Storing power for use when required makes very small sources of power much more useful. Lighting, for example, is a very important load, which is not continuous. When it is dark, in the evening, a household will use several lights for a few hours per day, then shut them off, go to bed and not use them again until the next evening. A battery system stores up power during the late night and through the day and allows outgo to exceed input for a while each day.

Motors in common use can require many times the amount of power that they need to run, just to get started. Inverters will produce the high output, referred to as "surge capacity" that is several times the continuous output for this purpose.

Most microhydro systems have more power than they can use, at least sometimes. When this is the case, the size of the inverter is by far the most important factor in system performance. A larger inverter will make a system seem much larger, especially when the loads to be considered are motor loads such as refrigerators and freezers.

In general, resistance heating and batteries do not mix. Therefore, items like clothes dryers and electric ranges, are probably not going to work satisfactorily on your battery charging system. If the inverter could start one of these loads, it is dubious that there would be enough kilowatt hours available in the batteries to run them for a useful length of time.

Alternators

A common practice is to charge batteries by generating AC with an alternator and then rectifying it, which means turning it into DC, to charge batteries.

Several different kinds of alternators are used. Automotive alternators are readily available and inexpensive. On the other hand, they are relatively inefficient. This means that the output from a given flow of water is lower than it could be, or if more water is available, that more water is required to produce a given level of service. Their brushes need periodic replacement. They might last a year or so under light duty, less under heavy load.

Some permanent magnet alternators are designed specifically for microhydro, and will adjust so that they are efficient at a wide range of rotating speeds. The permanent magnet design means that no regular maintenance is required. The life expectancy of bearings can be many years.

Induction motors can be used as brushless alternators as well. They may suffer from efficiency and control problems, but they are extremely inexpensive and rugged.

Oversizing alternators is a perfectly reasonable strategy; a 1-kilowatt brushless alternator will produce 25 watts with reasonable efficiency. At low power levels there is much less resistance, and as a result, the peak of efficiency of an alternator may be a small fraction of total output. This is a convenient fact, since one size can fit many sites. For example, the case studies "Upgrading a Small AC System" and "A Relatively High Output System" use a 12-kilowatt rated alternator, even though the Buck Creek system generated as little as 1.75 kilowatts, and the Relatively High Output system generated 10 kilowatts.

Turbines

For the purposes of battery-charging microhydro, "high head" means heads from 6 feet or so, up to 600 feet. Impulse turbines, such as the Turgo or Pelton, are used. There is some overlap with the range of "low head" reaction turbines used for heads between 2 feet and 10 feet. A run of the river turbine like the Aquair uses the current of the river instead of drop.

AC microhydro requires relatively higher head, at least 75 feet or so, to be considered high head. Getting multiple kilowatts of AC from heads lower than this is not often financially attractive.

Impulse Turbines for High Head

Impulse turbines are used for situations with heads ranging from 6 feet to 600 feet.

Turgo

In one commonly used impulse turbine, the Turgo, the jet of water strikes the runner at an angle. The photo (next page) shows a Turgo impulse turbine used especially for situations with high water flows. Because it can use more water, significant power can be generated with less head, which results in shorter penstocks. The wheel can use up to four 1-inch jets. It is rugged and can also be used as a small and inexpensive AC turbine for the right kind of site.

The Energy Systems and Design Stream Engine
Turgo turbine. Credit: Energy Systems and Design.

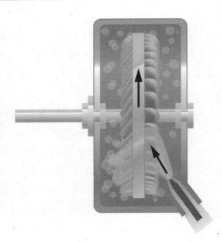

The Turgo turbine. Credit: Corri Loschuck.

The Harris Pelton runner.
Credit: Harris Hydroelectric.

Pelton

In another type of impulse turbine, the Pelton, the jet of water strikes the runner along its circumference. It can be slightly more efficient than the Turgo, and is used especially for low flow, high head situations.

One Pelton, from Harris Hydroelectric, will generate some power with as little as two gallons per minute, and it will accommodate up to four half-inch jets to handle more flow. It has a 4-inch pitch diameter, and thus will turn at 1,800 revolutions per minute with a pressure of only 32 pounds per square inch.

Table 4.1: Energy Systems & Design Stream Engine Output
(in continuous watts)

head in feet	flow in gallons per minute				
	10	20	40	75	100
10	–	20	50	90	120
20	15	40	100	180	230
50	45	110	230	450	600
100	80	200	500	940	1,600
200	150	400	900	1,600	–

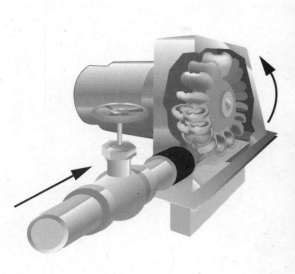

The Pelton turbine. Credit: Corri Loschuck.

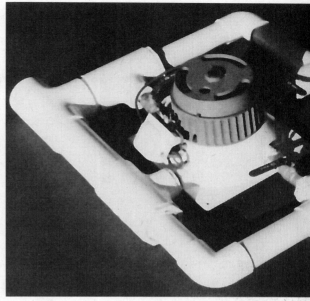

The Harris Pelton turbine.
Credit: Harris Hydroelectric.

Table 4.2: Harris Hydroelectric Permanent Magnet Alternator Output
(in continuous watts)

flow in gpm	net head in feet					
	25	50	75	100	200	300
3	–	–	–	–	45	80
6	–	–	30	45	130	180
10	–	40	75	95	210	300
15	25	75	110	150	320	450
20	40	100	160	240	480	600
30	65	150	250	350	650	940
50	130	265	420	600	1100	1500
100	230	500	750	1100	1500	–

Note how much less efficient the automotive alternator is with the same turbine.

Table 4.3: Harris Hydroelectric Ford Alternator Output
(in watts)

flow in gpm	net head in feet					
	5	50	75	100	200	300
3	–	–	–	–	30	70
6	–	–	25	35	100	150
10	–	35	60	80	180	275
15	20	60	95	130	260	400
20	30	80	130	200	400	550
30	50	125	210	290	580	850
50	115	230	350	500	950	1400
100	200	425	625	850	1500	–

Water Baby

The Water Baby is a new unit from Energy Systems and Design being introduced as this book is going to press. Neither a Pelton nor a Turgo design exactly, the tiny 2-inch diameter runner turns a small brushless alternator, with up to 4 quarter-inch jets.

The Water Baby in action.
Credit: Energy Systems and Design.

The Water Baby alternator and runner.
Credit: Energy Systems and Design.

The Water Baby is designed to work with very small flows of water. Since waterpower is very generous, as we have seen, small outputs are very useful. The Water Baby, given enough pressure, will produce the 25–300 watts that most sites actually use. Most of our case studies that charge batteries could easily use the Water Baby. In many ways, you can consider the Water Baby a "right sized" machine, no bigger than it has to be to deliver satisfaction. And, because it's smaller, it costs quite a bit less than other similar turbines.

Another feature of the Water Baby is the new permanent magnet brushless alternator. It is highly efficient over a wide range of speeds. It also can be adjusted while running, which is very convenient indeed.

Low Head Turbines

Low head for battery charging systems means heads under 10 or 12 feet. Turbines that work in this range are reaction type turbines, for example the LH-1000 also from Energy Systems and Design. It will produce power from as little as two feet of head, and is rated at 1,000 watts at 10 feet of head, using 1,000 gallons per minute.

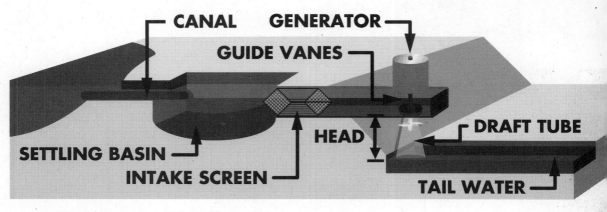

A typical low head system. Credit: Corri Loschuck.

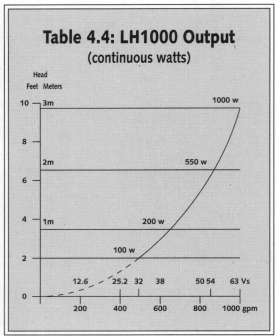

Table 4.4: LH1000 Output
(continuous watts)

Left: The ESD LH-1000 in action.
Credit: Energy Systems and Design.

Low Head AC Turbines

The new Powerpal technology from Asian Phoenix Resources can be used for sites where low cost is really important. Although designed as a complete AC system, the high voltage output and sine wave AC waveform make them ideal candidates to move small amounts of power to charge batteries a long way from the powerhouse at lowest expense. Since their output is 60 cycle AC, they can use off-the-shelf components to transform voltages to the high levels necessary for efficiency. A 240-volt model is available as well as the 120-volt model which is the standard for North America.

Table 4.5: Powerpal Performance
(in watts)

model	MHG-200W	MHG-500W	MHG-1000W
Water head	5 feet	5 feet	5 feet
Water flow	554 gpm	1110 gpm	2060 gpm
Output voltage	110 or 220 VAC	110 or 220 VAC	110 or 220 VAC
Output frequency	50–60 Hz	50–60 Hz	50–60 Hz
Output power	144 kwhr/month (200 watts continuous)	360 kwhr/month (500 watts continuous)	720 kwhr/month (1000 watts continuous)
Generator weight	35 pounds	70 pounds	165 pounds

Above: The Powerpal in action.
Credit: Asian Phoenix Resources, Ltd.
Left: The Powerpal turbine. Credit: Scott Davis

The Aquair turbine.
Credit: Jack Rabbit Marine.

Other Turbine Solutions

The Aquair is a marine product that can be used for microhydro, where site conditions are suitable. Its original use was to be towed behind a barge to provide navigation lights. It offers some opportunities, in that no penstock is required. On the other hand, it requires very high stream velocities, nine knots or so, which are hard to come by in fresh water streams. As with wind velocity, people chronically overestimate water speeds and so are disappointed by its performance. Just to compare, water velocities in penstocks rarely exceed five knots.

Because the turbine is just out there in the water, there is a significant chance of damage from debris. It does come with a spare prop however. It would be improved for stream use with another prop that operates at five knots or lower.

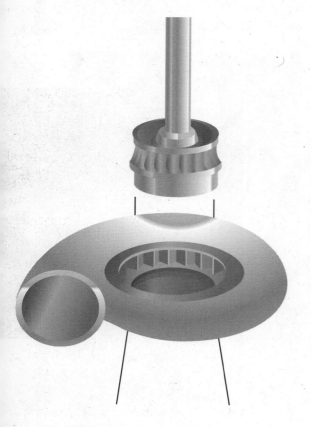

The Francis turbine looks like a centrifugal pump.
Credit: Corri Loschuck.

Constructing Your Own Turbine

A centrifugal pump is very much like a Francis turbine, only running backwards. Pumps are cheap and readily available. This may seem like an attractive alternative to high turbine prices, and it has been done successfully many times. There are, however, a few issues with this idea. The exact details of how pumps work as turbines are not commonly specified, and unless the particular pump you are using has been measured for output in the particular application of head and flow you are using, predicting its performance is guesswork. But, let me guess this: the efficiency will be quite low.

The output of the Francis turbine is quite sensitive to changes in head and flow. Basically, each size of turbine has a narrow range of optimum head and flow. This contrasts with the impulse type turbine, like the Pelton or the Turgo, which can operate efficiently at a wide range of heads and flows.

An induction motor, provided with excitation through a capacitor, will generate electricity when run backwards. Such an arrangement would provide brushless reliability without the cost of brushless alternators, and provide inexpensive small alternators.

An induction motor used as an alternator. Credit: Ann Cavanag

Like pumps running backwards as turbines, there are many issues with using induction motors as alternators. They are not notoriously efficient in this mode. If the speed that they run at is very different from their design speed, they won't work very well at all. In practice, since there are only a few different sizes of turbines available, this means that only certain heads will be appropriate. As load goes off them, the voltage rises, sometimes precipitately, which can damage equipment if not strictly controlled.

That said, induction motors can be used as alternators, for example, in the case study "A System with a Long Transmission Line." The reason this kind of alternator was used at the time was to make a high voltage system with a brushless alternator that would transmit power a long distance. The water pressure was in the correct range to ensure that the rotating speed of the turbine was high enough without being too high. In this system, power is generated as three phase AC, transmitted on a three wire system to the batteries, where it is transformed to the correct voltage and rectified. This method is very practical where transmission distances are long.

On the left is a hand made turbine that powered a 32-volt DC generator for many years. On the right, my father thought that this cooling fan would also make a waterwheel. Credit: Roy Davis .

A crossflow turbine can be fabricated with materials that are readily available in third world situations. They may offer the best solution for multiple kilowatt AC from lower heads.

In fact, spraying water under pressure at many things will cause them to turn and make power. (For samples of "homebrew" turbines, see <www.otherpower.com>.) Although it may seem a lot cheaper to fabricate your own, and I certainly have done so myself, it should be mentioned that low hydraulic efficiency is just like low electrical efficiency: you may be outsmarting yourself. Projects that use more water are larger projects, regardless of output. More water and larger pipes for a given output may increase overlooked costs like larger valves and plumbing. Think about it.

Balance of System Equipment

As well as the turbine and alternator, the rest of the system you need is called "balance of system" equipment.

Batteries

RV Batteries – Not Good Enough

Batteries with the "RV" designation are not true deep cycle batteries. They are a poor compromise — of the kind of that gives "compromising" a bad name — and offer some cycling performance while still being able to start an engine without damage. They do not deep cycle very well and will fail quite soon. "Deep cycling" means that batteries can be discharged deeply and recharged many times without reducing their life expectancy.

Forklift Batteries – More Than Required

Although the large, 2 volt cells, commonly called forklift batteries are excellent true deep cycle batteries, microhydro systems seldom need the capacity that this kind of battery provides.

Golf Cart Batteries – Usually Just Right

By contrast, the kind of batteries that are called "golf cart batteries" are made to deep cycle. They are the smallest batteries of this sort available. This is the kind that is most commonly used for microhydro systems, since large capacity is not required. Since they are not worked very hard, they last a long time.

Other Battery Technologies

Other battery technologies may be required. If they are to be flown into a site, safety considerations dictate a sealed battery such as a gel cell or the absorbent glass mat (AGM) technology.

Battery Charge Control

Microhydro turbines charge their batteries continuously. In all systems, sooner or later, the batteries get fully charged. They need to stop charging to prevent over charging. Microhydro is powerful enough that this situation may occur daily or more often. Even if the system is being worked hard, eventually you go away or something and the batteries have all the charge that they need.

Charge controlling technology is similar to but less demanding than governing an AC system. Power levels are lower, and the requirement is simply to prevent the batteries from overcharging by providing a load as required. By contrast, AC systems must precisely regulate the speed at which they rotate at all times.

When the batteries are charged, a diversion control device switches on a dump load. This load is a small heater of some sort, at least as large as the full output of the turbine, and not so large as to exceed the current carrying capacity of the controller. It can be an air heater, or a water heater, perhaps in the tail race. Often the powerhouse is de-iced with the waste heat.

While it is possible to capture this energy, the quantities of heat are often quite small. Remember, even a kilowatt of power has only 3413 BTU's. Since the dump circuit handles only the power that you produce and don't use, the amount of domestic hot water likely to be heated by a DC system is pretty small.

Diversion Loads

Another important issue is getting suitable loads for diversion loads. Since the voltage is low, even small outputs require large currents to operate. Special high current, low output devices must be found or fabricated for diversion loads.

Inverters

Inverters make battery power into the AC power that we are all familiar with. An off-grid site, such as a cabin with a Capacity Self-rating of one or so, can get good service from a 500 watt inverter (or even smaller).

If your requirements are about a two or so on the scale, a 1.5-kilowatt inverter powers a remarkable range of small tools and appliances, while a 2,400 watt or larger capacity inverter will provide modern conveniences, rated about three on the capacity rating scale. A 4-kilowatt sine wave inverter will provide a remarkable level of high quality power for a larger household or light commercial application.

Small AC Systems

If you rate your needs at one, two, or three on the Consumption Self-rating Scale, this chapter will explain how it is also possible that a small AC system, a kilowatt or smaller, may meet them.

For many years, conventional wisdom had it that an output of three kilowatts or even higher was required for an AC-only system. This perception was based partly on fact — a microhydro system generally has the same output all the time (the continuous rating and the peak output are the same). However, electrical usage goes up and down, sometimes abruptly. Since many kinds of motors in common use may use up to ten or more times their running power requirements to get started, it was felt that power levels of two kilowatts or more were minimal to provide this service.

However, this is not necessarily true. Many of the issues that require large amount of power, such as a starting surge, can be worked around successfully. For example, a capacitor start kit can be installed on refrigerators to lower their startup requirements. After all, two continuous kilowatts delivers far more kilowatt hours than most on-grid homes use. A little efficiency means that far fewer kilowatt hours are required than you might think.

Generating AC power is inherently simpler and more efficient than generating AC, rectifying it to DC, charging batteries, and using an inverter.

Just to be fair, though, it must be said that some people like the idea of having a supply of energy stored up. That way, if the water is shut off for some reason, there is still some power available. It can boil down to a matter of taste.

Another factor that makes AC generation attractive is that the output is sine wave in form, just like the power that comes out of our plug ins. As long as the frequency is held stable by a governor, sine wave AC doesn't buzz and motors run cooler and more efficiently. Sine wave inverters give better service than the modified sine wave inverters, but are more expensive.

The last factor that makes AC generation at any power level interesting is that it is generated at higher voltage than DC usually is. This makes for increased efficiency when transmitting the power any distance. Also, readily available transformers can be used which cost much less than custom transformers.

After all, what do you really need? North Americans demand a lot from their renewable energy systems. For rural electrification in developing countries, and in many off grid situations here, a few lights and a bit of power for other things is entirely worthwhile. If you use the formula "One person, one light; plus one for the room" and consider that a 16 watt compact fluorescent is an excellent light, you can see that power outputs of 1,000, 500, or even 200 watts will make a tremendous contribution to life and productivity. The case study "A Low Head AC System" makes this point dramatically.

To sum up: on one hand, small AC systems are extremely simple and efficient. Unlike many battery charging systems, their output is sine wave, and motors run cooler and more efficiently with this kind of power. Lighting, in many cases the most important single use of a microhydro system, fits this kind of system very well.

On the other hand, the peak output of a small AC system is no higher than its average output. This means that a system like this does not have very good performance when it comes to starting motors or doing anything that briefly requires a lot of power.

Governing Small AC Systems

In general, the problem with governing AC systems is spinning the alternator at its rated speed, rather than keeping batteries at a certain voltage, as is the case with governing DC systems. Most alternators are

Table 4.6: Weight of Pipe and Water
(in pounds per foot)

type of pipe	lbs/foot	type of pipe	lbs/foot
Schedule 40 PVC		Poly	
2"	.07	1.25"	.38
4"	2.05	1.5"	.45
6"	3.61	2"	.62
8"	5.45	–	–

designed to run with diesel powerplants and are rated at 1,800 rpm. Some run at even slower speeds like 1,200 or 900 rpm.

With a diesel powerplant, when the load varies, more or less fuel is injected into the cylinders and so the speed remains constant for the alternator. Fuel injection can change fast enough to deal with the kind of changes in electrical load that are experienced. Regulation provides stable voltage.

In the hydroelectric situation, the moving column of water rather than the combustion of fuel provides the energy. In order to change the power output, the flow of water must change. This is entirely possible but expensive, difficult, and not as good as throttling fuel. Water generally does not move very fast in a pipe (perhaps a brisk walking pace at best), but it is very heavy.

This means that there is considerable inertia in the water as it flows. If you stop it or even slow it down suddenly, this inertia expresses itself as "water hammer." Considerable forces are generated, which can break pipe and vibrate fittings loose. Electrical loads cut off almost instantly; however, water's flow rate doesn't change as instantly, and so the two match with difficulty. The voltage regulation technology that comes with the alternator does, however, tend to smooth out consequences of this effect.

In the past, a spear valve was placed in the jet of water and actuated by a hydraulic actuator. The controller, the Woodwards UJ8, for example, sensed the rotating speed and adjusted the spear valve accordingly.

This setup worked pretty well, but was prohibitively expensive in the eyes of many potential developers of small sites. The power quality was not as tightly controlled for voltage and frequency as grid power. Luckily, this was not an important consideration for many essential loads.

Electrical systems require governing because the loads they serve typically vary over time. In the beginning, hydroelectric plants only ran at night to provide lighting. Without governing, alternator speed, and thus frequency and voltage, would change rapidly with load. Even though a turbine might be able to function at a variety of speeds, the same is not true of the alternators used in AC systems. They must operate within certain narrow limits to keep the AC frequency stable, and especially cannot be allowed to rotate very much faster than they are rated to

This system required little governing.
Credit: Scott Davis.

spin. AC alternators are easily damaged from overspeed. Loads can also be adversely affected by high (or low) voltage or frequency. Governing is essential.

Until the advent of electronic load control, governing technology could easily be the most expensive and problematic part of a microhydro system. This is a major reason why many sites were not developed years ago.

Luckily, AC systems do not require perfect governing for many kinds of loads, and even a little power is very useful.

The strategy in the case study "A Small AC System" was to trade potential output for controllability. Generally, to extract the power in water most effectively, the turbine must turn at a speed which is about half that of the jet of water that strikes it. This is accomplished by gearing the turbine and alternator together so that the load on the alternator will slow the two down to the proper speed. In this scenario, a governor prevents a sudden drop in load from allowing the turbine and alternator to turn at nearly the speed of the jet of water instead of half. This would be, in severe cases, like doubling the speed of the alternator.

Because AC alternators are designed with a constant speed prime mover like a diesel in mind, they risk damage when operated at speeds higher than those they are designed for.

The success of this experiment hinged upon important features of the impulse turbine. Reaction turbines are very sensitive to operating outside of the envelope for which they are designed. Not so with the impulse wheel, as we shall see.

The fastest that the turbine would ever turn with the pressure available was about 600 rpm with no load whatsoever. Since the turbine turned the alternator with a belt drive, the choice of sizes of sheaves determined the speed at which the alternator would run. A large sheave on the turbine turned the smallest usable sheave on the alternator at about a 3.1:1 ratio.

Here's how it worked out in practice: Some small load was left on all the time. A fluorescent light that seemed immune to high voltage worked for years. Then, as more loads came on, the speed of the alternator gradually dropped. However, the more the alternator slowed, the more efficient the system became, and so there was a kind of stabilizing effect there. Voltage regulation meant that even if the frequency was reduced proportionally to the load, the voltage dropped more slowly. This meant that the lights would indeed vary in brightness with load. More lights on meant dimmer lights for all.

Looking on the bright side however, the system had good rotating inertia. To start electrical motors, a surge current is required which may be many times the amount required to operate it. This surge only lasts for a second or two in the case of a refrigerator, and the energy for it came from the rotating inertia.

This is an excellent example of the exception proving the rule, where the rule is that AC energy cannot be stored. A small yet significant amount of energy stored as inertia improved the surge performance quite a bit.

The penstock could theoretically deliver more than two thousand kilowatt hours per month. In this configuration, it delivered about 435 kilowatt hours per month. Is this good news or bad news?

An important thing to remember is that the first hundred kilowatt-hours per month is the most important. This amount of power will deliver the most essential services to a remote household, like lighting. And so, in many ways, the output from this system was generous.

This setup delivered at minimum initial cost, and little subsequent expense, nearly half as much power as an average house in town. Off grid refrigeration is an achievement and, while it is now possible to buy custom refrigeration that is very, very good, this system ran a very inexpensive refrigerator for many years, even though the voltage was forever at 90 volts. Lights are dimmer at that voltage, but bulbs certainly last a long time.

Remember the seventies were a time when the inverter was not very practical for remote use. They were available, but had many problems in actual use. Even though relatively expensive, the performance left a lot to be desired. The signal was not a sine wave, such as comes from a rotating object like a alternator, but a square wave, which is noisy and inefficient. Inverters had reliability problems in that they were so expensive that they were regularly undersized, and that led to premature failure as they were routinely overloaded.

More Early Practices

Another solution to governing small systems used at the case study "A Small AC System" involved a voltage-sensitive solid state relay connected by a dimmer to a hot water tank element. This combination had three knobs to twiddle, one for high voltage dropoff, one for low voltage pickup, and one knob to adjust the size of the dump load, for best performance. It worked rather well for a couple of years, until it was replaced by a modern electronic load control system.

Modern Solutions

Very small AC systems don't necessarily need a lot of governing, especially if they are used primarily for lighting, which is the sort of a load that is very forgiving of odd frequencies and voltages.

However, these smaller systems will definitely benefit from sophisticated governing systems such as electronic load controls. This technology will ensure that the output goes where it is needed most, and that all output goes somewhere. It provides output with a stable frequency, which increases its usefulness. Not everything is as forgiving as a light bulb. This makes the most of a limited resource, and can be very valuable.

AC Systems

If you rate yourself much over a three on the Consumption Self-rating Scale, or a four on the Capacity Self-rating Scale, you will require a larger AC system. As mentioned earlier, there is considerable difference of opinion about precisely where the dividing line is between battery charging systems and AC systems. Traditionally, larger systems are AC and smaller systems charge batteries. If the continuous output of a system is high enough to meet your needs for surging capacity, no battery/inverter subsystem is required, and AC can be generated directly.

We found that an AC system started motors at least as well, if not better than an inverter of the same rating. For example, the 2.5-kilowatt system on page 121 could start and run a Maytag washer that the modified sine wave 2.5-kilowatt inverter at the "System with a Long Pipeline" could not.

Should your potential output fall between a couple of hundred and a couple of thousand continuous watts, there is some choice of technologies. There is no doubt that modern inverter technology makes small amounts of power very useful by allowing the electrical output to match the load. If there is a large load for a few seconds, the system will carry it. This has been standard practice in North America for many years.

Some people just hate the idea of batteries, and for those people, the AC system is best. An AC system produces a sine wave signal that is the same kind as you get from a plug in, and so appliances of all sorts like it the best. Sine wave inverters are available, but they can easily be the most expensive portion of a system. Even though the surge capacity of an AC system is low, there are ways to work around this problem if desired. The AC system is simpler than a battery and inverter system, with fewer parts.

Since AC systems include systems that are larger than most of their DC counterparts, some heating is often a part of the services offered by an AC microhydro system. A 10-kilowatt system that runs in the winter provides heat that is the equivalent of burning 12 cords of fir firewood in a six month heating season.

I have had years of operating experience with AC systems having outputs from about 600 watts of continuous output on up to 3.2 kilowatts. Some clients had systems that were much larger. Even with the smallest system, we got excellent lighting service, even in the winter, and it did start and run a very ordinary electric refrigerator.

The system was upgraded and when the output of the system reached about 1,000 watts, it began to produce a significant amount of domestic hot water as a dump load. As the number of kilowatt hours per month reached about the same as we use in town, the amount of hot water available rose as well. Left overnight, the excess power would ensure that there was hot water in the morning.

Since, however, there were only a thousand watts available, minus whatever we used elsewhere, the recovery time on the hot water tank was pretty lengthy. If you used up all the hot water, it took a long time to come back. The water tank was acting like a battery and storing up energy.

Upgrades continued to the system and each time the power output increased, the recovery performance increased also. At about 1,750 watts continuous output, we found that we began to get some space heating as well as domestic hot water. At about 3,000 watts, a significant amount of the space heating load was taken up by the hydroelectric system, as well as significantly improving the hot water tank recovery time.

Efficiency

Is efficiency necessary, where power is too cheap to meter? In a microhydro system, after the initial investment is made, the ongoing costs are often negligible. You may have to replace drive belts every year or two. Brushes for alternators only cost a few dollars. Many systems don't even use drive belts and brushes. Intakes may fill with debris and require cleaning, damage may occur, but these expenses are nothing like buying fuel for a generator. Essentially, the fuel is free.

So, is efficiency necessary? First, no system is big enough to support waste. Once, I attended an off-grid site where a 30-kilowatt system was supporting eleven households. This means that each place got about twice the kilowatt hours of an average North American household. Yet, the system was still "too small," in that consumption regularly exceeded capacity. Interestingly enough, a mile down the road a system that produced 0.2-kilowatts for a household (that is, less than ten percent of the per capita power level that they used just next door) was operated with success for many years. The difference was motivation.

I had this experience with the Buck Creek system: Over the years, I took a small system and did many different things to increase the power output:

- Larger nozzles (which meant more intake maintenance)
- Different pulleys to improve efficiency
- More rotating inertia
- Switching to a direct drive system to eliminate belts and increase efficiency

The list went on and on; since we were our own power providers, the more power the merrier. During the heating season, more than six months a year, every last watt went somewhere useful. Because it was an AC system, the small output we had in the beginning meant that there were serious limitations on how much total capacity was available.

Conservation is a most powerful technology, but there are limits. Energy efficient lighting should be everywhere, but when it comes to other appliances, you should consider whether the money would be better spent improving the capabilities of the system as a whole, or being more efficient.

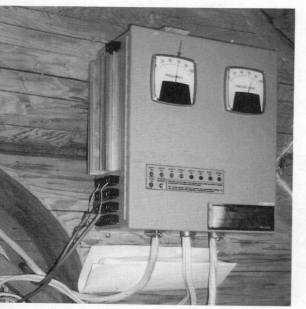

Thompson and Howe electronic load controller.
Credit: Scott Davis.

While the power potential from a PV system might be quite limited by budget, the choice of a more efficient turbine and alternator may solve many, many problems, as we discovered in the case study "Upgrading a Small AC System."

Governing Larger AC Systems

Load control in AC systems is similar in some ways to charge control in battery charging systems. A constant load is maintained on the alternator at all times. However, in battery charging systems, batteries provide most of this load and determine the voltage of the turbine. Only when batteries are fully charged does the diversion controller operate to prevent overcharging.

In an AC system, frequency is a critical parameter. The frequency of the output is determined by the rotating speed of the alternator. For AC systems, this governing function is called "load control" and operates continuously to keep variations in the frequency down to a half a cycle per second regardless of load.

Frequency is monitored in the load controller and kept within specifications by adding or shedding dump loads in a two step process. Coarse adjustment is done by a series of prioritized relays. These allow any type of load to be used as a dump, as long as it can be randomly switched on and off. Space heaters are commonly used on relay steps.

Here, excess power not used for water heating or space heating is dumped through baseboard heaters on the back porch.

Here, excess power is dumped into space heaters on the back porch. All systems have excess power sometimes. Credit: Scott Davis.

Fine adjustment is done with a triac circuit. This circuit works like a dimmer or motor speed control, except that it is controlled by the microprocessor in the load controller unit to adjust the load on the system so that the frequency remains stable.

Not all loads can be hooked up to the triac. Resistive loads such as heaters are most suitable. It will not function as a speed control for a motor. Typical dump loads for triacs are domestic hot water heater elements. A load controller makes sure that all the output of an AC system goes somewhere useful, as well as providing close frequency regulation.

AC Turbines

The smallest AC systems are the Powerpal series. They produce 200, 500, and 1,000 watts. They are complete units that include a controller.

For outputs over a kilowatt or two, AC is used exclusively. Most AC microhydro systems in this size range use impulse wheels, like the Pelton or the Turgo.

In fact, in the case study on page 121, a turbine which was designed for battery charging systems, the Energy Systems and Design 4-inch Turgo, was found to be plenty sturdy enough to be used in AC systems up to four kilowatts or so. The system was also featured in the 1999 *Canadian Renewable Energy Guide.*

Unlike the adjustable permanent magnet DC alternators, AC alternators must turn at their rated speed to produce the proper frequency as well as voltage and current. This speed is most often 1,800 rpm, and so there are many different turbines in use for AC generation. This high speed is easiest to get from small diameters. A 4-inch turbine, for example, turns an 1,800 rpm alternator optimally at about 80 feet of head, whereas a 7-inch turbine is optimized at about 300 feet of head.

Belt Drives

Conventional alternators are readily available. However, they do have to rotate at their rated speed, usually 1,800 rpm. Many times, alternators and turbines are coupled with belts instead of by direct drive in order to deliver the best performance from both. Usually, a turbine that has a smaller diameter will be geared to an alternator so that a less expensive turbine can be used.

However, every site offers unique elements. The case study "Upgrading a Small AC System," had a turbine that was larger in diameter than required. These big antique wheels were originally designed for processes that required low speeds, such as grinding or running a sawmill. They can be used for generating electricity, but proper gearing is required.

Belts add maintenance, since they require replacement and occasional tightening.

Low Head, Multiple-kilowatt AC

This kind of system is possible but unlikely. Turbines are priced more by their size than the output they produce. A large volume of water flow means a large turbine. Turbines that handle many thousands of gallons per minute are large and so expensive that they often make this kind of project impractical financially. (However, I have seen photos of a site using up to a dozen of the Powerpal type of turbine.)

Transmitting AC Power

AC is easily generated at high voltages. Readily available transformers can step up the voltage even higher to minimize power losses over long distances. However, since power outputs in AC systems are higher than in battery charging systems, the current carried may be much higher even though the power is being generated at higher voltages. Conductor costs can be very high for AC systems.

Long AC power lines have another problem. To review the math: losses are proportional to the square of the current. This means that effects of surges in demand on performance are much magnified by long transmission lines with high losses. Such power lines limit the surge capacity of a system and are only practical when charging batteries, where surging is not a problem.

5

Getting Started

THIS CHAPTER EXPLAINS HOW TO GET GOING ON DEVELOPING your
microhydro resource, including what kinds of water power resources you
need, the equipment that can provide it, and how to install the various
elements into a hard working system that will last for many years.

Specifically, you need:

- At least two gallons per minute of water flow, and a lot of drop.
  – or, two feet of drop and 500 gallons per minute of water flow.
  – or some combination of head and flow that is in between these
  values whose product is a useful amount of energy. Heads of up to
  600 feet can be practical.

- The proper turbine and alternator.

- A powerhouse to keep the turbine out of the weather.

- This resource needs to be generated within a kilometer or two of the
  point of use to be cost effective.

- Permission, from the relevant authorities, even if the project is
  entirely on your own land. Penstocks may extend onto public land,
  with permission. A narrow right of way is commonly given.

- An intake, to get the water to where it is to be used. Water exits the
  creek, loses large debris in a stilling basin, is screened for smaller
  debris, and enters the power system.

- Pipeline or penstock, if you have a high head system, or a flume or a
  ditch or some way of moving water. The penstock is often the most
  expensive and difficult part of a larger microhydro installation.

– If you are using pipe, use a vacuum breaker to prevent collapsing the intake screen or worse yet, the penstock, in an emergency. Here's why: If the top end of the penstock is blocked, or if the plumbing breaks near the bottom, water rushing out of the pipe will leave a vacuum in the penstock. This means that air pressure, which is 14.7 pounds per square inch, can be pressing on the pipe. Penstocks can be collapsed this way. I saw this work in a small way at Buck Creek. We had a hose connected to a hose bib at the house. From the house, the far end of the pipe was way down hill. If you shut the hose bib off while the other end was open, the hose would squish flat from vacuum every time.

– A pressure gauge for monitoring. But remember, do not install a valve on the upper end of the penstock without a working vacuum breaker mechanism.

• A transmission line to move the power that you generate from the alternator to the point of use. This line should be short, if possible. It can, however, be a mile or more away and still be cost effective. Penstocks are most often buried for protection from frost and damage. This can be a lot of work and expense. A short penstock, all things being equal, is best.

• You need to be able to turn the turbine off, or to dry out the penstock in case repairs involving gluing pipe are required. This means ideally a valve of some sort at both ends of the penstock.

In addition, battery charging systems need:

• Batteries, if you have a small site, to store power for best service.

• Diversion controller and load, in battery charging systems, to keep the batteries properly charged.

• Inverter to make AC from the batteries. Usually, AC appliances are cheaper, better, and more readily available than DC ones.

AC systems need load control or other governing. Diversion loads can be common appliances such as hot water heaters or baseboard heaters. The smallest AC systems come with their own controller. Larger AC systems consist of turbine, alternator and load controller.

About Pipe

Choosing the correct pipe for the job is essential. Many different combinations of pipe diameter and length will create power. I have been quite amazed at the high level of service that electronics brings to a couple of hundred yards of poly pipe.

Polyethylene (Poly)

For microhydro, 2-inch poly pipe is perfect. Unless freezing is an issue, a microhydro system can be as easy as unrolling a couple of hundred yards of poly pipe and plumbing a turbine to the end. When burying pipe, poly requires little by way of bedding for protection during backfilling. Given enough head and flow, and the proper hydroelectric system, it will produce far more power than any affordable number of photovoltaic panels, and do so on cloudy days and all night as well.

Additionally, 2-inch poly pipe is a very common size to deliver winterized water to a household. It is large enough to provide some fire protection. At ordinary household pressures, it is large enough to generate a significant amount of power, even though it might be thousands of feet long.

Even smaller pipe has also been used for power generation and, given enough pressure, can perform surprisingly well. The case study "A System with a Long Pipeline" uses a couple of thousand feet of 1 1/4-inch pipe. Even 1-inch poly can have enough pressure applied to it to produce some power.

Poly pipe fittings are secured with hose clamps, using up to three clamps as the pressure climbs.

PVC

Another kind of pipe in common use is PVC pipe. I have also been surprised by how much power can be produced from a very steep 4-inch PVC pipe.

PVC is light and easy to carry.
Credit: Scott Davis.

PVC has low frictional losses, is relatively inexpensive, light to carry for installation and is used almost exclusively when the pipe is over two inches in size. Small systems, with 1.5 or 2-inch pipe, may use PVC as well. PVC comes in 20 foot lengths which are joined together by gluing, or with gaskets where the pipe is to be buried or otherwise restrained.

Sizing Pipe

Pipe size is determined by two things: performance requirements, based on the length of pipeline required and the optimum flow rate, and budget. A pipe that is too small will limit power production. A pipe that is too large will take resources that would have been better spent elsewhere in the system.

Penstocks need to be a certain length in order to get the drop they do. For example, in the case study "A System With A Long Pipeline," it took about 2,000 feet of pipe to get the 150 feet or so of drop.

The drop and the volume together determine the power output. Using two-inch pipe instead of one-and-a-half-inch pipe would have meant that more power would have been produced from the same length of penstock. It would have cost several hundred dollars more, but it would have been a cost effective way to make more power.

For every combination of penstock length and water flow rate, there will be a pipe size that will give a head loss of about a third. This is the smallest and cheapest pipe to use for the job.

If your budget permits, going to the next larger pipe size can be an effective way to increase performance from a given set of circumstances. Frictional losses can never be eliminated, but it is not impossible to get a 20 percent increase in actual power output by using the same amount of water and pressure with pipe that is the next size larger. It will increase the pipe budget by a third or even a half, plus seriously inflate the price of fittings, but it may be financially attractive. Just figure it out. Use up to date pricing, since pipe prices are quite volatile.

Low Pressure and High Pressure Pipe

A pipeline has low pressure at the top and high pressure at the bottom. Low pressure pipe is light, inexpensive and the easiest to carry around. For example PVC that is rated at 63 psi only weighs 50 pounds or so for an 8-inch diameter, 20 foot long piece. It is, therefore, perfectly acceptable to use 63 psi pipe until the expected pressure is close to the rated pressure, then switch to progressively higher pressure ratings toward the bottom of the penstock. We did this at the case study "A Relatively High Output System."

Microhydro Opportunities

Some site factors can make life easier for microhydro developers.

Spring Water

Spring water is generally warm and is thus easier to winterize. It may be that the water is warm enough that burial is not necessary, since flowing water will keep pipes from freezing down to 14°F (-10°C) or lower.

Existing Pipes

Existing pipes in a gravity fed domestic water supply system should be examined, since the pipeline is the most difficult part of a microhydro project. Even 1-inch pipe has significant power potential with high pressure, as long as the length is very short. The advantages are that they are already installed and winterized. The fact that a microhydro system runs all the time may also be of value to a gravity water system. The water is constantly fresh and the fact that it is kept moving will help to keep it from freezing.

Low Head Systems

A low head system should be investigated if you have a large enough volume of water, about 500 gallons per minute or more, again to avoid pipeline construction. Since only a small amount of head is required, there may be many different places to put such a system in a given landscape. This will give more scope to look for places that are easy to construct and maintain.

   In any case, ease of construction and maintenance should be considered at least as important as maximizing output. Each site is unique, so be sure that you are aware of all alternatives.

...even a couple of
amps is way better
than nothing!

Remember, even a couple of amps is way better than nothing! Outputs that seem pretty small can do amazing things because they get to store up their energy over time.

A system with an output of 50 watts, or 35 kilowatt hours per month, will give better performance than PV systems costing many times as much, especially in the winter months.

An output of 100 watts (75 kilowatt hours per month) is generous for delivering essential loads like lighting and electronics.

Three hundred or four hundred watts, about two or three hundred kilowatt hours per month, is a level of power that can be transmitted a long way, thousands of feet, without making the transmission line the major cost. And this power level gives great service, especially with a large inverter.

And, as you will see in the case study, "A Low Head AC System," whole villages can be run from a kilowatt of power for the very basics.

Microhydro Obstacles

Like any other civil engineering project, microhydro development, and especially the task of burying penstocks and constructing intakes, is subject to all sorts of different site conditions.

Very Steep Ground

Sometimes a site looks really good because the hillside is very steep, and the water is falling rapidly. As long as pipe is not to be buried, the main problem is securing the pipe from shifting. Small pipe like 2-inch poly can be tied to trees or boulders. Larger pipe may require the construction of thrust blocks, which are masses of concrete cast in place around the pipe to prevent the pipe from moving.

If pipe is to be buried, very steep ground may indicate that the hillside is rocky with a thin layer of soil over it. This was the situation at the case study "A Relatively High Output AC system" site. Luckily, the pipe was buried mostly for physical protection from sun and impact rather than protection from the mild winters, and so a couple of feet of soil was enough for the job. As it happened, the excavator could reach soil from both sides of the pipeline for cover, which was fortunate, because the soil was quite thin.

Plant grass on disturbed soil to prevent erosion after digging.

Waterfalls

Waterfalls look good, but the same problems that occur with very steep ground are present at waterfalls. Waterfalls generally fall over rock, or else they would have long ago eroded into a series of rapids. This makes pipe burial very difficult.

Debris

Some creeks carry an enormous amount of sediment and debris in high water. This can bury or carry away intake structures. In a situation like this, it may be that service cannot be provided during periods of high water. However, many clever solutions have evolved to deal with problems like this in the thousands of years that people have been taking water from creeks and rivers.

See what other people in your neighbourhood do to solve this problem. Remember, when it comes to ideas, "good artists borrow, great artists steal."

High Water, Low Water

Varying flows affect many water sources. In the case study "A Small AC System," a dozen sandbags diverted water into the intake channel or back into the creek as required. This was enough to handle, at this site, up to 800 gallons per minute of irrigation and hydroelectric water.

As your plumber, let me advise you to handle as little water as possible. This is an important reason why efficiency remains important, even though the power is "too cheap to meter." At the Buck Creek ranch, the intake was adequate for irrigation and a small flow for hydro, but as the flow rate through the turbine went from about 150 gallons per minute to over 500 gallons per minute (plus more than 200 gallons for irrigation in the summer), intake maintenance increased proportionally.

As your plumber, let me advise you to handle as little water as possible.

Freezing Weather

A bit of weather below freezing may not affect microhydro very much. Unless the weather continues freezing hard for many days, the moving water will keep 2-inch poly on the surface from freezing.

When winters are protracted, the soil freezes. First the surface, then as freezing weather continues, the depth that the soil is frozen increases. In order to keep pipes from freezing, they would have to buried deeper than the frost depth. Although the severity of the winter is a good indicator of frost depth, snow does insulate. A resort in the alpine zone of the Monashees could bury their penstock by hand to minimize environmental impact, because the soil never froze underneath all that snow.

Ice

Icy conditions certainly pose problems for microhydro development, but then again, people have been bringing water from one place to another a lot longer than hydroelectricity has been around, and so we have learned a few things.

Ice, unlike the solid form of practically anything else in the universe, is actually lighter than the liquid form, water. Because of this, ice floats. Ice is actually a fairly efficient insulator and so water continues to flow underneath the ice.

However, waterfalls and even rapids can entrain enough frigid air in the water that they can freeze right to the bottom. Thus it is important to have a stilling basin that keeps the water relatively smooth. In the case study "A Small AC System," a long ditch brought water from the creek to the microhydro intake. The water was a couple of feet deep in the ditch portion, and about four feet deep in the stilling basin portion. Even though the ditch ran on the dark and cold side of the mountain for many hundreds of feet, it continued to run winter after winter underneath the ice.

Installing Microhydro Equipment

Installing a microhydro system is a civil engineering task, but you can use this book as a guide to doing it yourself.

You must divert water from a stream or gather it from a spring or other water source, remove debris, and deliver it to the turbine.

In a high head system, the penstock is a pipeline that delivers water to the turbine. It is often buried for protection from the elements. The turbine and alternator are mounted at the high pressure end of the pipeline. The powerhouse needs to provide a sturdy mounting platform, protection from the elements, and a way for the water to escape the wheel and flow away freely in all weather.

In a low head system, the water is delivered to the turbine at no pressure, and head is developed after the turbine in the draft tube. The powerhouse is at the intake. The draft tube must deliver the five or ten feet of head required. It can be vertical, or it may slant depending upon site conditions. In any case, the outlet end must be below water level. A flume may bring water to the turbine.

After electricity is generated, it must be transmitted to where it is being used. Electrical transmission lines may be buried, which is very safe and well protected from accidents. Power lines may be overhead as well. This, when done correctly, is also safe and less expensive. They are more exposed to accidents however. Avoid running conductors on the surface of the ground for fire and other safety concerns. Even a small amount of power can be dangerous.

Penstocks

Penstocks are often buried for protection. PVC deteriorates in the sun. Physical protection, from falling trees or rock, for example, is provided by burial which also eliminates the hazard of the pipe shifting around while water is running. If the pipe is buried below the frost line, it will be protected from freezing and can be expected to run in the winter.

Even if the pipe is not quite below frost line, the fact that the water is running through the pipe may help to keep it from freezing.

Frost depth can be six feet or more in some climates. Credit: Scott Davis.

Although PVC is quite tough and resilient, it can be damaged as it is being buried by jagged rocks underneath or being dumped on it. The solution is to "bed" the pipe before the trench is backfilled. In an ideal situation, sand would cover the pipe a few inches deep to cushion against the impact of heavy rocks, but ideal situations may hard to come by in remote areas.

Sometimes pipe isn't buried at all.
Credit: Scott Davis.

Generally, a person with a shovel and a rock rake will smooth the bottom of the trench as necessary, glue the pipe together, and shovel some material around the penstock to cushion it during the backfill process. Even an inch or two of material on top of the pipe will protect it from harm.

Alternatively, instead of backfilling by simply pushing the fill back into the trench, it can be shoveled back in by machine in such a way that it does not drop directly onto the pipe. Instead, new fill is added where it has already been filled in to some extent. The new fill then rolls or flows into place. A person with a shovel can deflect big ugly rocks from crashing onto or resting directly on the pipe.

Excavators can bury 1,500 feet of pipe per day under most conditions. They will have to be specially prepared for work in a creek, involving such precautions as arriving clean to the worksite and changing to non-toxic hydraulic fluid. Let the regulations be your guide.

To sum up: most hydroelectric systems develop their power by water being handled in a pipe. They carry water from the intake to the turbine in a penstock. This means a pipeline project. If the system is to be winterized, the penstock can easily be the most expensive and certainly the most difficult part of any microhydro installation

It may not be necessary to bury your penstock. You may live in an area where freezing weather is not a problem. Remember, your penstock is running all the time. In order to freeze a running water line, extended periods of weather quite a bit below freezing are required. If you only require three season service, or you live in a mild climate where frosts are mild and of short duration, you will be able to avoid a most difficult part of microhydro development, the pipeline project.

When a penstock is not buried, it will need to be secured at intervals to keep it from shifting around. A thrust block is a concrete mass cast around pipe for this purpose. Thrust blocks are also a good idea where forces are high in buried lines, such as where they emerge from the ground to the powerhouse. I have also seen 2-inch poly tied to trees with success.

Plumbing

The penstock is likely to be the part of the system that requires the most work. As plumbing chores go, however, it is not too difficult. One pipe takes water from the intake to the turbine.

In addition to the pipe, certain fittings are required.

Coupling Poly

Poly pipe uses insert couplings and hose clamps for coupling. More than one clamp can be used for high pressure applications. A rubber hammer helps drive in couplings when they are cold. It may be necessary to heat up larger diameter pipe with boiling water, or very, very carefully with a torch.

Coupling PVC

Pipes are coupled together by either gluing or they may have ends with gaskets in them so that they are just shoved together.

To glue PVC pipe, make sure the pieces are clean and dry. This may not be easy, but persevere. There are a variety of glues and formats depending upon requirements. The main difference between glues is the speed with which they dry. Cold weather makes glues set more slowly. I have been snowed on lots while gluing. Try to avoid this. Some glue, especially for smaller pipe, requires just one step to apply. Other glues use a primer first to soften the pipe for the glue. With these, use primer on both pieces to be glued, then apply the glue over the primer.

Smaller sizes may use the applicator in the can. A paint roller can be used to apply the primer and glue in larger pipes. Quickly insert the pipe into the socket and turn a quarter turn to distribute the glue. Hold a minute or so until it is set.

Glue comes in several speeds for use in a wide variety of temperatures. A fast glue sets more quickly in low temperatures. Let the glue dry overnight before pressurizing the pipe for maximum strength. Work fast, since the glue can set up quickly in hot weather. Always protect your hands with rubber gloves when working with glue.

Coupling Gasketted Pipe

Gasketted pipe is coupled together by lubricating the gaskets with special potable lubricant and shoving the pipe ends together. A hint: they only go together when they are lined up accurately. Even then, a little delicate force may be required for larger sizes. Once I had to use a sledge hammer to get them started. I used a wooden driver to distribute the force around the end of the pipe and was very, very careful. The case study "A Relatively High Output System," used gasketted pipe.

Above: First lay out the pipe.
Credit: Bonnie Mae Newsmall.
Below: Then, connect it up. That's me in the circle.
Credit: Bonnie Mae Newsmall.

Valves

You need a valve to turn the turbine off. Servicing is seldom required, but may occasionally prove necessary. If you have one jet, a valve near the jet will be satisfactory. If you install a valve on each jet, then the system does not need to be sized for the minimum flow of the creek. Jets can be turned off one at a time and so a wide variety of flows can be accommodated.

Pressure Gauge

Every system needs a pressure gauge. It will help you tell when things are going well and what is wrong if they are not. Usually, the pressure will be constant.

High Pressure

If the pressure reading is too high, most likely the nozzle is clogged. Something has gotten past the screen and traveled down the pipe and lodged in the nozzle. This happens sometimes and is why it is best to make sure that the nozzle can be removed for cleaning if necessary.

Low Pressure

If the pressure reading is too low, most likely there is not enough water being delivered to the intake. Another way of looking at this is that the jet is too big for the water available. Rather than running a full pipe, the level of the water in the pipe has gone part way down the penstock. This can create quite a powerful suction on the intake and pipe. Avoid this situation by switching to a smaller nozzle.

Vacuum Breaker

Sometimes an accident will result in all the water rushing out of the penstock, without more water rushing in. Maybe the intake gets suddenly clogged, or the penstock gets broken, or something. Should something like this happen it is possible that the suddenly empty penstock will collapse from air pressure because it is under vacuum rather than pressure. A vacuum breaker is installed to let air into the pipe when it drains, even if the intake is clogged. Vacuum breakers can also be a simple tee in the penstock, with an air pipe that goes above the level of water in the intake. Protect the open end to make sure nothing falls in. Then, air can enter the pipe and prevent accidents. 1-inch pipe is good enough for 2-inch pipe; a 1.5-inch pipe is good for up to a 6-inch pipe; and a 2-inch pipe will do for 8-inch and larger.

Plumbing in the Powerhouse

Reduce pressure losses by using the largest pipe and fittings you can, right up close to the jets. Turbines should be protected from the weather.

Freezing in the powerhouse is seldom a problem because the alternator produces some heat that tends to keep the place dry and thawed. Just make sure there is no leaky plumbing! Flexible pipe may also be used to deliver water to a jet so that jets and valves are easier to change.

Sizing Pipe

Water is *not* a gas — keep this in mind. Since water is virtually incompressible, reducing the size of the pipe does not increase the pressure or do much of anything except add to the frictional losses and cost you power. Use the largest pipe and fittings you can justify financially. Keep small diameter sections of pipe short.

Use large diameter pipe right up to the valve to reduce losses. Credit: Energy Systems and Design.

However, high pressure pipe is much cheaper in smaller diameters. It is perfectly acceptable to use large diameter pipe at low pressure and then switch to smaller diameters when pressures climb lower down in the penstock. Do a careful analysis of the factors of cost and difficulty versus power lost or gained by various scenarios.

Intakes

Each microhydro system has an intake, where the water enters the penstock from its natural source. Any debris in the water is removed.

Easy to say, perhaps not so easy to do. But on the positive side, people have been getting water out of springs, creeks, and rivers for a very long time now, and there is considerable experience built up.

Streams vary widely in their characteristics. I operated a hydroelectric system with a less than optimum intake system, and it still didn't require much maintenance because the creek was easy to manage. A dozen sandbags moved around to deal with high water ran a system that used up to 800 gallons per minute between the irrigation and the microhydro system.

Other systems on other creeks nearby had more trouble because the creek was full of gravel in the spring. High water was very high. In situations like this, choosing an optimum site, where there is protection for the intake works, might take precedence over getting the maximum amount of head.

Costs reflect the volume of water used, not the amount of power produced. This may be a drawback to certain potential low head projects. The intake must take water out of the creek, provide a stilling basin, and deliver it to the forebay. The forebay secures the top of the penstock and holds the screen in place.

Stilling Basin

Most debris in streams and creeks either floats near the surface or sinks right to the bottom, given a chance to settle. A stilling basin slows water down so that the debris can drop out of it before getting to your intake screen.

Some of the debris consists of large heavy pieces of gravel and rocks. This material will settle out if the water is allowed to level out and slow down in a stilling basin. The larger the amount of water used, the larger these facilities must be. The stilling basin must be flushed out or dug out periodically as it fills up with gravel. An overflow pulls off the debris that floats and delivers it back to the creek. Ideally, the intake screen is submerged beneath the frost line. It is housed in a protective forebay.

Stilling basin and forebay. Credit: Scott Davis.

95

Intake Screens

Conventional screening methods use a durable screen material such as stainless steel or brass. The mesh size should be small enough to block anything from entering the pipe that would clog up the nozzles.

Sloping sides make for self cleaning.
Credit: Scott Davis.

Choosing screen size carefully is important. Screen that is too fine, such as window screen, clogs up immediately. I found that quarter-inch mesh screen was a good compromise between being small enough to be effective and large enough not to require constant maintenance.

Screens should be made of tough and durable materials. Stainless steel screen is available in agricultural irrigation outlets.

Large screen areas are desirable, since they will take longer to clog up. A rule of thumb is to provide a square yard of screen area for each cubic foot per second of water flow. This is quite a lot of area. The screen is often wrapped around a form to be more compact.

Remember, many microhydro systems use sealed bearings you don't have to lubricate, and utilize direct drives, which means that you never need to change a belt. Cleaning the intake is only maintenance requirement. Thus the intake is worth doing well.

Conventional screens should be as close to vertical as possible or at least slanted so that debris has a tendency to fall off. A horizontal screen has no tendency to self clean. If possible, arrange for the water that is not used in the penstock to flow across the screens to encourage self cleaning.

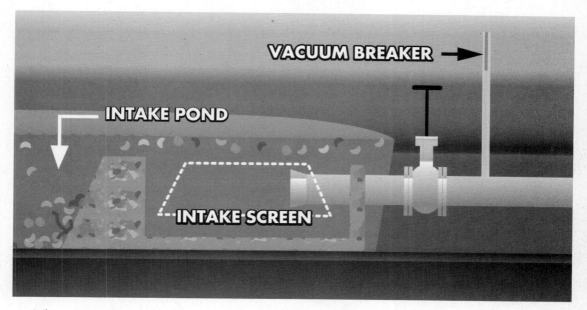

There is a screening system, based on the Coanda effect, which is self cleaning. This product is used to take the water out of coal slurry. Surface tension does the trick. There is a great article in *Home Power* magazine (June/July 1999) on this topic.

The intake screening should be angled to encourage self cleaning. Credit: Corri Loschuck.

6

Regulations and Incentives

THIS CHAPTER OFFERS ADVICE ON THE LEGAL AND FINANCIAL SIDE of microhydro development, the "carrots" and the "sticks" that can make a project go...or not.

The Carrot: Government Incentives for Microhydro

Appropriate Incentives for Appropriate Technology

Of course, accounting to the "triple bottom line," where ecological and social costs are acknowledged as well as the more narrowly defined financial costs, would be the most appropriate incentive for appropriate technology. If we were actually responsible financially for the consequences of our energy behavior, no particular "programs" would be necessary for renewables.

Luckily, we don't even have to wait for this collective "coming to our senses" to happen. Over decades, I have had abundant experiences proving that off-grid, where the power lines do not go, renewable energy is totally cost effective. Just think of power lines as delivering subsidies, too.

The ideal incentive for renewable energy development would be simple and easily accessible to those who need it. It would continue to provide motivation to keep the equipment running throughout the lifetime of the project.

Long term low interest financing is one such incentive. Even where no incentives exist specifically for renewable energy, microhydro equipment has long qualified for farm financing. Check your local situation.

Rebates are a common type of incentive, which can be effective if the product costs just a few dollars (for example, an energy efficient light bulb).

However, as the price of the equipment increases, several problems with rebates as incentives occur. Renewable energy on-grid is essentially energy-saving technology, and the market penetration of energy-saving technologies is governed by the "Payback Acceptance Curve." Basically, if the payback period for a technology is five years or more, the market penetration will be low. If the payback is a year or less, the market penetration is very high.

Rebates have to be very generous indeed to overcome low energy prices. By contrast, it is easy enough for the payments on a low interest, long term loan to be lower than energy costs. The payback time is zero.

Rebates do not offer continued incentive to keep a system running, only to buy it. This can have the effect of attracting companies whose main expertise is selling, rather than installing or maintaining equipment. Then, as systems do not operate as advertised, the misconception that renewables "don't work" is created, when in fact it was the incentive that didn't work. Again, long term financing encourages good system maintenance in order to keep making the payments.Net metering is another incentive, by offering retail rates to producers (in the form of credit). This technology is well worked out and is really an indicator of willingness on the part of utilities to cooperate with renewables.

Independent power production, where the generation of renewable energy is encouraged at all scales from the smallest on up, is in fact the important issue here.

Cogeneration, where a factory or mill generates its own power, say from burning waste to make steam to generate electricity, and uses the waste heat from this process for its own heating needs, is an idea that has been blocked for reasons of policy for many years.

Incentives come and go as well, so it is best to find out about your local incentives when the time comes; interesting and up-to-date references on incentives can be found online or through your local authorities.

Incentives in Canada

As this book goes to press, British Columbia has begun to investigate providing net metering. Currently, the Northwest Territories offers — through the RETCAP program — 15-year, prime interest rate financing for renewable energy projects, including microhydro.

Manitoba offers net metering, however extremely low energy prices make it less attractive.

For specific details on incentives in Canada, see Friends of Renewable Energy BC's study of Canadian renewable energy incentives at <www.forebc.com>.

Incentives in the US

As of the time of publication, 35 US states have some form of net metering, and more are undoubtedly on the way.

For a detailed breakdown of other US incentives, see <www.desirusa.org>.

The Stick: Regulations

Hydroelectric generation is a non consumptive use of water, and in the past I have found government agencies helpful and the requirements reasonable. However, recently I have found this to be less so. Regulations and policies come and go, so be sure to investigate your particular circumstances with the appropriate authorities. Remind them, if necessary, that properly done, microhydro will generate clean power for many years without damaging the environment in any way.

The Process in British Columbia

When you apply for a water license, the Water Management Branch will contact the Department of Fisheries and Oceans (DFO). DFO will provide a referral letter for the project called a "letter of advice." When you apply for a water license from the Ministry of Land, Water and Air, you must also apply for a "license to cut" permit from the Ministry of Forests if you are working on Crown Land where there are trees that have to come down for penstock construction. Sometimes these timber values can be considerable.

Department of Fisheries and Oceans

Water is routinely removed from watercourses without interfering with fish, and has been for many years. Screening is the main ongoing requirement, and the screening required to keep fish fry out of the pipe is about the same as you would use to prevent damage to turbine runners anyway. When constructing an intake, it is necessary to follow certain guidelines so that a simple incident like a broken hydraulic line doesn't mean a fish kill.

The Department of Fisheries and Oceans produces documents which are very helpful on the topic of intake screening.

Department of Water, Land and Air Protection

Water rights can be easy to come by, as long as there aren't any competing uses or other difficulty. Water rights are prioritized by date of application in case any conflicts arise.

It is usually quite possible to get a narrow right of way through public land if necessary for penstock construction. Paying stumpage fees may also give you the timber on the right of way.

Of course, no project on Crown Lands will happen where Native land claims and treaty rights are being actively contested.

7
Case Studies

WHAT FOLLOWS ARE ACTUAL EXAMPLES of working microhydro systems that have given good service over the years. Each one illustrates important features that fit together to make a successful system.

Just a reminder — getting good service from a system means it is doing the job it was intended to do. For example, in remote areas there may be no utilities available. More than one technology may be used to deliver service, much as in a town where you might heat and cook with gas, and use electricity for other things.

One task that is often asked of microhydro in remote areas is to provide enough electrical power to run those essentials that can't really be run with other technologies. While propane refrigerators are a well worked out technology, we have yet to run a laptop on bottled gas. Yes, propane can also power lights but they are neither cheap to buy, cheap to run, nor entirely satisfactory in a number of other ways.

Microhydro, PV and wind can all charge batteries, and as such have a lot in common. Any technology that relies on energy storage in batteries will be limited in the amount of heating it will do. However, electricity can provide essential services, those services that only electrical power can provide, at remarkably low power levels with conservation and high efficiency technology.

Electric lighting is the best kind. Efficient lighting means that supplying this kind of service doesn't have to take a lot of power. Good, essential service ("lights and music"), for an off-grid household can be supplied, whether with wind, PV, or microhydro, or some combination, with a few dozen kilowatt hours per month. Prospective microhydro developers should realize what small levels of continuous power are actually required for a high level of service.

Since microhydro is such a powerful technology, many potential users can lose sight of the fact that solar and wind sites have made a science of more from less. These same efficient technologies and practices make very low levels of microhydro output even more significant.

The thing that makes microhydro powerful is that water generally flows all the time. By contrast, the wind blows at various strengths and then doesn't blow at all. The sun goes behind clouds, down every night and south for the winter. Also, microhydro systems don't usually use the whole water supply available. The pipe size required to confine even a small creek can be prohibitively expensive and may be unnecessary.

For example, an energy efficient bulb might use 16 watts. In the dark time of the winter you might have it on half the day. A single irrigation sprinkler uses water pressures and volumes that could light such a bulb.

In all cases, it is important to figure out what service is required and to size your system appropriately. It may take less than you think.

These case studies range from small to large, casual to elaborate. What they have in common is that they work.

The first group of case studies are of smaller systems, where microhydro acts like other renewable technologies to charge batteries.

Case Study 1:
A System with a Long Pipeline

The Problem

Years ago, a 2,000 foot piece of 1 1/4-inch poly pipe was buried to provide winterized domestic water to a homestead miles from grid power outside Gold Bridge, BC. Although the stream is small, the pressure is good and there is excess water at the intake. The owner had questions: Could they run a turbine off the water in the pipe to produce enough electricity to run lights, radio, radiophone and stereo? Would they still have enough water to run showers and toilets? All they wanted was about a 1 or 2 on the Consumption Self-rating Scale (page 18), and so they chose to get an inverter that would provide a rating of about 2 on the Capacity Self-rating Scale (page 25) as well.

Table 7.1: About a System with a Long Pipeline

Pipe length and material	2,000 feet, 1.25" polyethylene
Level of service desired	"lights and music"
Water volume available	lots
Static pressure	65 psi
Net pressure	50 psi (100 feet of net head)
Flow rate	7.5 gpm (gallons per minute)
Turbine	Harris Hydroelectric with Ford alternator
Output	50 kWh/month (70 continuous watts)
Transmission distance	25 feet
Voltage	12 volts
Consumption Rating	2
Capacity Rating	2

The Solution

The static pressure of the water in the buried pipe was measured (with a gauge attached to a convenient hose bib) at 65 pounds per square inch.

A 12 volt Harris/Ford Hydroelectric unit was installed close to the house to minimize transmission losses and cost. A couple of golf cart batteries and a 500 watt inverter made up the battery/inverter subsystem.

The ideal water flow volume for this length and diameter of pipe is about seven or eight gallons per minute. A 7/32-inch nozzle gives about this flow at a net head of 50 or so. Note that this still leaves a lot of pressure available for sprinklers, running water from the faucets, and the like.

The system delivers 50 kilowatt hours per month of continuous, clean, free power all year long. It would take a 300-watt peak PV array to average this amount of power. Three hundred watts of PV costs at least 50 percent more than the turbine and requires many times more battery storage. Even then, winter performance would require another source of power such as a wind turbine or fossil fuel generator.

This system provides plenty of power for lights, radio, radiotelephone, and stereo. Other services like refrigeration and cooking are provided with propane.

A System With A Long Pipeline. Credit: Corri Loschuck.

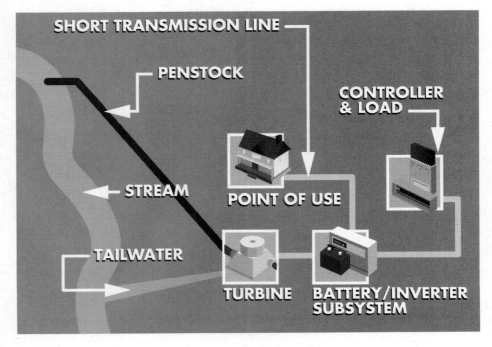

Replacing This System Today

Let's assume that the waterline is in place, as it was when we came upon this project. It would cost about the same price as a small motorcycle, $2,100–$2,400 (Cdn$3,500 to Cdn$4,000) plus labour to replace this system today. A photovoltaic system in this price range would produce half or so of the output of this system on the average, and much less during the critical lighting period of the winter. Remember, these people do not need to use a fossil fuel generator for lighting backup in the winter.

Water striking the turbine runner. Credit: Scott Davis.

Case Study 2:
A System with a Long Transmission Line

The Problem

These musicians live in a lovely place, far from town. It has a small stream, one that may or may not muster enough volume to reach the culvert underneath the road below. Even so, it provides irrigation and domestic water to a household. Power could be generated about 1,700 feet from where it would be stored and used. The owner had these questions: Is it too far away? Can it power electronics like a music studio?

Table 7.2: About a System with a Long Transmission Line	
Penstock material and length	3-inch PVC, 500 feet long
Flow rate	20 gpm
Transmission distance	1700 feet
Output	125 kWh/month
Turbine	Early *ESD plastic turgo with induction alternator
Inverter A	2.5 kW
Inverter B	600 watts
Consumption Rating	3
Capacity Rating	4
	*Energy Systems and Design

The Solution

The homestead is more than a quarter of a mile as the crow flies from the creek. An irrigation system was installed to water the garden and a few acres of pasture. An intake was placed in the creek, and pipe was laid to the pasture. Buried for frost protection, it provided domestic water to the house, as well.

Part way along, a tee in the line brought water down a steep hillside to the road. This became the penstock.

In this site, an induction motor is used as an alternator. Electricity is generated at 240 volts for transmission 1,700 feet to the site where the power is being used. At a power shed, it is transformed to 12 volts and rectified to DC where it charges batteries.

When I interviewed an owner, the system had been running for a decade or so. He had forgotten how much head there was, since a gauge hadn't been installed or a convenient place included to put one. The jet size was also unavailable, since it had run for so many years without maintenance.

However, I would estimate that the flow is about 20 gallons per minute or less, and the static head is about 150 feet. Since the 3-inch pipe is oversized for this flow rate, losses are negligible.

The induction alternator system is not very efficient. The long transmission line and homebrew transformers created some losses as well and so the power delivered to the point of use is about 170 watts or about 125 kilowatt hours per month. This is lots of power for lights, refrigeration and the music studio, and so it never seemed necessary to the owner to optimize the system: "It just runs and runs." Two inverters power the place. One, a 2.5-kilowatt modified sine wave inverter, provides high capacity at low cost. A second inverter, a 600 watt sine wave unit, was added to provide the high quality sine wave power required to run a music studio, which it does flawlessly.

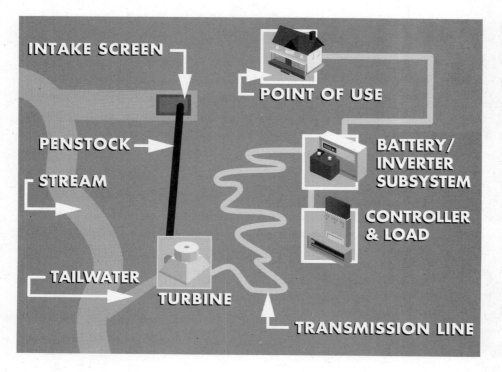

A system with a long transmission line. Credit: Corri Loschuck.

A system with a long transmission line.
Credit: Ann Cavanagh.

Music as well as lights! Credit: Scott Davis.

Replacing This System Today

If this system were commissioned today, rather than many years ago, several aspects of the system would be done differently to make more from less. For example, using the more efficient technology available today, much more power could be produced. A high voltage brushless alternator and turbine, together with a transformer/rectifier pack would be used. Eight golf cart style batteries would be specified instead of the forklift style batteries that were available when the system was assembled. At today's prices, it is possible to get a sine wave 2.5-kilowatt inverter that gives much better performance for nearly the same price as the two inverter system.

As well as higher efficiency, a multijet system could produce many more kilowatt hours per year by using more water in the seasons when it is available.

At today's prices, this system would cost about the same as a big ATV, approximately (Cdn$7,100), plus tax and installation. About a third of the total cost would be for the turbine/alternator, and about half would be the inverter/battery subsystem. The transmission line and a proper transformer/rectifier unit would cost only about ten percent of the total cost.

Case Study 3:
A Simple System

The Problem

The problem was to provide power for a remote household where a northern winter required pipe burial below frost depth, but without the attendant bother, expense and permanence. With a family of five, a full-service household was required, including electrical refrigeration as well as lighting, tools, and other appliances.

Table 7.3: About a Simple System

Penstock length and material	600 feet, 2" polypropylene
Flow rate	100 gpm
Nozzle size	1 inch
Turbine	Energy Sysytems and Design Stream Engine, a 4" bronze turgo with brushless Alternator
Static head	100 feet
Net pressure	15 psi
Power output	150 kWh/month
Consumption Rating	3.5
Capacity Rating	4

The Solution

The major construction consideration for microhydro in cold climates is protection of the pipe from freezing temperature. Burying below the frost line is the usual method, giving excellent mechanical protection as well. However, in rugged terrain, this can be a formidable task.

Two-inch poly is terribly rugged. Here, the owner just unreeled the pipe in the creek bed to protect it from freezing. This is not recommended, as holes can become abraded into the pipe over time as it rubs against the rocks. However, this system has run for a few winters without insurmountable problems, so there you go.

This system has to count somewhere in the three's on the Consumption Self-rating Scale, and a four or so on the Capacity Self-rating Scale.

The owner tells us that there is about 600 feet of pipe laid in the creek bed. A pressure gauge shows that there is about 100 feet of static head. The net pressure is about 15 pounds. Since the owner told us that a 1-inch nozzle is used, the water flow is about 100 gpm.

The system delivers 150 kilowatt hours per month to the batteries, a few hundred feet away. In conjunction with a 2.5-kilowatt inverter, it runs an ordinary refrigerator/freezer as well as giving plenty of lighting and other electronics for a family of five.

This illustrates that electrical generation can be both pretty casual and quite powerful. Actually, this system has a nozzle that is too big for the length of the pipe used. More water is being used than is strictly necessary, meaning that the frictional losses are higher than they should be.

However, since there is lots of water in the creek, and lots of power in the batteries for the loads that are being used, it just wasn't worth resizing the nozzle, says the owner. Besides, lots of water flow keeps it from freezing up.

This system delivers modern conveniences without the sophistication of a sine wave inverter. It would rate as a three on the Consumption Self-rating Scale and a fthree on the Capacity Self-rating Scale.

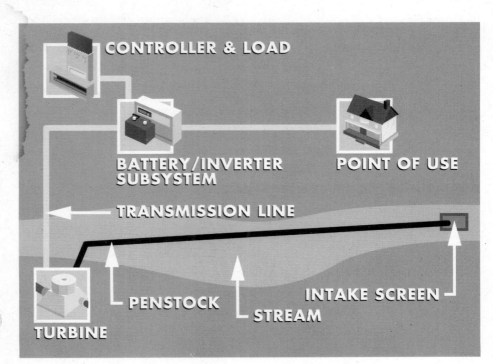

A simple microhydro system. Credit: Corri Loschuck.

Replacing This System Today

This system produces about three times as much power and gives service that is definitely a step above that of the case study "A System With A Long Pipeline." It also has the potential for using the water more efficiently. The turbine/alternator unit requires no periodic brush replacement. The inverter and battery subsystem are much more powerful.

It would, however, cost about twice as much as the "System With A Long Pipeline" case study system, more like a medium sized motorcycle, about $4,100 (Cdn$6,800).

A simple system. Credit: Scott Davis.

Case Study 4: A Small AC System

The Problem

A ranch that has never had grid power has buried pipe to supply an irrigation system and to provide domestic water to the house during a Canadian winter. The pipeline is steep and not very long. Although money is short, ingenuity is not.

The Solution

Buck Creek Ranch is a long way indeed from utility power. It is watered by Buck Creek, a small stream at one time professionally estimated have a minimum flow rate of about 750 gallons per minute. It falls many thousands of feet down the mountain behind the ranch, at about a 20 percent grade.

This is steep, but not so steep as to be rocky and impossible to dig. In the early '70s, a sprinkler irrigation system was being developed, and part of this effort saw 5-inch PVC pipe, 500 feet long, buried from an old mining ditch off the creek to the farm. It provided plenty of pressure for irrigation, and would give winterized water to the house.

Buck Creek Ranch. Credit: Susan Brown.

The First Solution — "Buck Creek A"

In the mid seventies, I helped to survey Buck Creek for a village scale hydroelectric system. The plan was to generate as much electricity as was practical and transmit it a few miles to where it was to be used.

We had some steel pipe that, although used, was still rated for 300 psi. This meant that the maximum head was about 660 feet. At a 20 percent grade, that means a penstock 3,300 feet long.

A flow of about 750 gallons per minute would occur at 440 feet of net head with an 11/16 inch jet. The turbine had a pitch diameter of 24 inches and would turn at

1,100 rpm and be belted to drive a 900 rpm alternator. The expected output was in excess of 70 kilowatts. The output would be shared between Buck Creek Ranch and its neighbour a couple of miles away via a high voltage three phase transmission line.

The Second Solution — "Buck Creek B"

"Buck Creek A " was a large project. Equipment was being assembled as financing progressed, but there were many delays. This particular project was shelved for many years, but was commissioned in a much different form in 2001.

In the meantime, the Buck Creek ranch owner borrowed the turbine that was intended for this project and plumbed it into his existing buried irrigation line. He installed the largest pulley he could afford in order to get the alternator speed up a little. It was never correct, in the sense of conforming to conventional wisdom. Much potential power was lost. However, this system worked very well.

Table 7.4: About a Small AC System: Buck Creek B

Penstock length and material	500 feet, 5" PVC, plus 12' 2" poly to jet.
Static pressure	49 psi
Net pressure	45 psi
Turbine	Vintage 1906 Pelton
Alternator	1,800 rpm, 120 volt AC alternator
Output	600 watts AC
Consumption Rating	3.0
Capacity Rating	3.0 – 4.0

The success of this experiment illustrates important features of impulse turbines, those turbines where a jet of water strikes a runner. It show that this kind of turbine is flexible about the flow and head of water that it uses.

This particular machine was an antique Pelton that had seen hard use. The cups were worn and hardly in top condition, although the shaft was straight and the runner was uncracked.

Since this was for ranch use, the idea was to spend no money. In the history of AC power development, governing has been the difficult and

1983: Many hands make lights work. The original Pelton generated five to seven kilowatts for many years. Credit: Roy Davis.

expensive part. Governing could easily be the largest single expense in a small system. These issues were dealt with at Buck Creek in an innovative and quite successful way.

The fate of this original Pelton was interesting. As if to illustrate the flexibility of the impulse turbine, this machine:

- was chosen for an extremely high head situation.
- was used in a situation where the head and flow were much lower.
- came to rest at last at a third place with intermediate pressure and output: 100 psi, where it has been generating about five kilowatts for many years.

The Third Solution — "Buck Creek C"

Our family bought the Buck Creek ranch in 1987. Although another turbine took the place of the original Pelton, the setup was much the same. It gave great performance all winter — far better than our neighbours got with photovoltaics — gave sine wave performance, and ran a very ordinary electric refrigerator. You'd rate this system on the Consumption Self-rating Scale (page 18), somewhere in the very, very low threes, and the Capacity Self-rating Scale (page 25), in the high threes.

Table 7.5: About a Small AC System: Buck Creek C

Penstock length and material	500 feet, 5" PVC, plus 12' 2" poly to jet.
Static pressure	49 psi
Net pressure	38 psi
Turbine	24" diameter turbine of unknown origin
Alternator	1,800 rpm, 120 volt AC
Output	1.1 kW AC
Consumption Rating	3.8
Capacity Rating	3.0 – 4.0

This "new" wheel at Buck Creek was interesting. It had such an odd design that a cup was pried off the wheel and taken to a manufacturer to identify. The conclusion was that it was a remarkably bad design, scrapped perhaps for its inefficiency....

Nevertheless, there was still lots of water left in the creek that didn't go down the pipe.

When we arrived, I installed a pressure gauge on the system. Always install a pressure gauge, because of the interesting things you can learn.

In this case, the net pressure was 45 pounds. Considering that the static pressure was 49 pounds, the net pressure could be as low as 32.3 psi. This, and our observation that we used only a small fraction of the creek flow, led us to believe that there was considerable untapped potential in the existing pipe.

I replaced the nozzle with a larger one and the working pressure dropped accordingly. As the working pressure dropped, the ratio between the turbine speed and the alternator became even more awkward.

To try to deal with this, I took the flywheel from a hay baler, a massive thing weighing a couple of hundred pounds, and had a groove machined in the perimeter for the drive belts. This made a large diameter sheave that

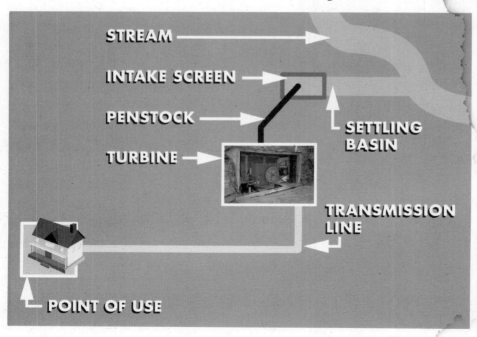

A Small AC System
Credit:
Corri Loschuck.

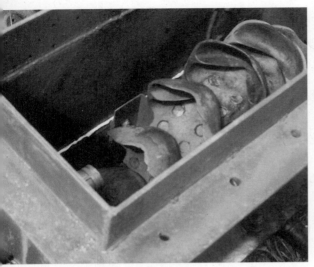

Despite the odd shape of the cups, this turbine ran for many years. Credit: Scott Davis.

Buck Creek C. Credit: Roy Davis.

acted as a flywheel and increased the rotating inertia many times.

All this made a significant increase, about 50 percent, in the amount of power available. Governing became a bigger issue.

If a large load is kept on all the time an AC system tends to be self governing. If only a small portion of the load changes, then the speed only changes proportionately; that is to say, not much. This method of control is often practiced for a while until people grow tired of it. It requires turning something on whenever something is turned off. Another inexpensive plan is to never turn off the lights. It has been done.

However, electrical power is terribly useful and as long as you have the plumbing, continued development of the resource has its rewards.

Our first actual controller (other than ourselves) was a solid state relay with adjustable voltage pickup and drop-off points. It was connected to a small hot water heater element acting as a dump load. The dump load was adjustable with a household light dimmer. In practice, the three knobs were occasionally tweaked to provide more starting surge or more stable operation. Fluctuating frequencies can make electronic equipment act oddly.

This was cheap and pretty effective. It enabled the addition of a freezer and automatic washing machine to the domestic scene and we got some (not a lot) of hot water from about a 1,000 continuous watts of output.

The best improvement to this small system, however, in terms of bang per buck, was the use of energy efficient bulbs. It was easy, too. Buy a couple of hundred dollars worth of bulbs. Get stepladder. Screw in. This alone made visible improvements in the capacity to start motors, and in the general behaviour of the system.

Lighting is in many cases the most important service for remote power systems to supply. If instead of using 500 watts for lighting, it now took us 100 watts, this was exactly like increasing the power output and capacity by 400 watts.

Case Study 5:
Upgrading a Small AC System

The Problem

The problem with the existing equipment at Buck Creek was that it was neither very reliable nor very efficient. We were using only a small fraction of the energy available from our site, and the pipe was already there.

The Solution

One Solution – Buck Creek "D"

As we continued to live at Buck Creek, we began to see that water was flowing by in the creek anyway, and that more potential could be developed "without too much trouble."

For example, even though there were up to thirty acres under irrigation, there was little pressure loss in the pipeline, five pounds or so. This meant, for the various nozzles that were tried, a small loss in total power output from the reduced pressure. Up to 320 gallons per minute came through a

Upgrading a small AC system. Credit: Corri Loschuck.

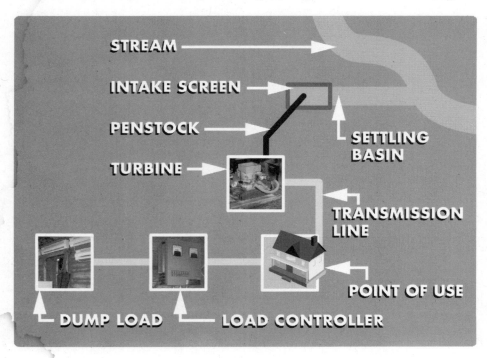

Table 7.6: Upgrading a Small AC System: Buck Creek D

Penstock length and material	5" PVC, 500 feet, plus 12' 4" manifolding
Static pressure	49 psi
Net pressure	40 psi
Flow	378 gpm
Nozzle size	2 × 1 inch
Output	2500 watts
Turbine	4" ESD Turgo
Alternator	12 kW brushless AC
Consumption Rating	4.2
Capacity Rating	4

1.25-inch nozzle, without running out of creek or interfering too much with sprinkler pressure, or being very demanding generally.

However, every kilowatt hour is worth a bit of firewood. Simply by running more water through an existing pipe, we could provide abundant domestic hot water, significant space heating in the winter, and music with the well behaved AC that would result. All we would have to do is boost the output to two or three kilowatts or more.

A more competent design would also eliminate powerhouse maintenance, which included at that time bearing greasing, belt maintenance and replacement and brush replacement. No one of these chores was particularly onerous, but collectively they did add up to a certain amount of work. Also, things have a way of breaking when it is not convenient for you to fix them.

In order to increase output and reliability we did two things:

- **Use more water**: To run more water through the penstock, we improved the plumbing from the penstock to the wheel. Originally, a 2-inch piece of pipe was used to get the water the last 12 feet or so from the penstock to the wheel. When flow rates were low (100 or 200 gallons per minute), this problem was more theoretical than actual. Yes, power was lost, but there were many more pressing problems, such as reliability and the rickety intake to worry about. Remember, the first watt is the best watt.

In order to get more usable power from higher flow rates, we replaced this 2-inch section with 4-inch PVC. This was a lot of work: thrust blocks, cutting through tree roots, roaring diesels, everything...it was actually "some trouble and expense," but it was definitely successful.

- **Increase efficiency**: We used a 4-inch bronze Turgo from Energy Systems and Design, which we mounted directly onto the shaft of a 12 kilowatt brushless alternator. This gave us higher efficiency as well as greatly improved reliability and reduced maintenance.

Although this turbine runner was designed for battery charging systems of a kilowatt or less, it is plenty strong enough for small AC systems up to four kilowatts as well.

Even though the 12 kilowatt alternator was larger than the system required, it was appropriate because it was available, brushless, and gave extra rotating inertia. The actual efficiency was high.

This new setup also increased the reliability of the system. Many times afterwards, I commented that the reliability was the best part of the whole upgrade. Many thanks to Appropriate Energy Systems!

This system was getting into a Consumption Self-rating Scale score of five or so and a Capacity Self-rating of four.

Buck Creek D with 4-inch Turgo turbine runner from Energy Systems and Design. Credit: Scott Davis.

The Next Solution – Buck Creek "E"

If our static head was 49 psi, then our net head could be as low as 32 or 33 psi. Since using two 1-inch nozzles did not in fact reduce the pressure to this level, we were not developing the entire potential output of the pipe.

Even with the irrigation running, the pressure was still higher than our calculated optimum net head and so we added a third nozzle to produce over two thousand kilowatt hours per month.

When do you quit? From Table 7.7, you will notice that the net head is even still not down to 32 pounds per square inch. Yet the capacity in the

Table 7.7: About a Small AC System: Buck Creek E

Penstock length and material	5" PVC, 500 feet, plus 12' 4" manifolding
Static pressure	49 psi
Net pressure	35 psi
Flow	528 gpm
Nozzle number and size	3 × 1 inch
Output	3,200 watts
Turbine	4" ESD Turgo
Alternator	12 kW brushless AC
Consumption Rating	4.5
Capacity Rating	5

pipe and stream to run yet a fourth nozzle was available. However, the volume we could run through the rickety intake was limited. A new and

better intake would have made it possible to run more water through the system. This was low on our priority list for a few reasons — we already had pretty good performance for water heating in the summer, and considerable space heating.

The turbine handles up to 200 gallons efficiently. We were using two and a half times as much water and the efficiency was actually dropping off a bit.

More power would not give us that much more service. We had reached a point of diminishing returns, unless we found something to do with the extra power.

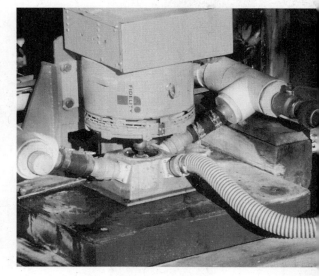

Buck Creek E. Credit: Scott Davis.

Replacing This System Today

This system would cost about the same as purchasing a small car today, perhaps $8,000 (Cdn$13,500), not including labor. The turbine/alternator unit and controller is about $4,200 (Cdn$7,000). Wiring and plumbing would be another $900 (Cdn$1,500), and the penstock and intake would cost about $3,000 (Cdn$5,000) to replace. This is about the same as a relatively high quality PV/wind hybrid system, and delivers up to ten times as much power.

As we began to use more water, intake issues began to be important and we spent more time fooling around with the intake. A maintenance schedule that was once a month became once a week. Once, the flow of the creek went down until it would only support two 3/4-inch jets, for a period of weeks.

But we did have a dishwasher and an outdoor soaker tub.

Table 7.8: Turbine History at Buck Creek

Turbine Version:	B	C[1]	D[2]	E[3]
Flow in gpm	113	320	378	528
Nozzle size in inches x number of nozzles	.75"× 1	1.25"× 1	1"× 2	1"× 3
Net pressure	45	38	40.5	34.6
Output kWh/month	425	800	1,800	2,300
Output/month in continuous watts	600	1,100	2,500	3,200

[1] = Bigger nozzle and flywheel, some load control
[2] = More water, more efficient turbine and alternator
[3] = Even more water

Case Study 6:
A Low-head AC System

A village on the Philippine island of Negros had no utilities of any sort, but it did have abundant low head hydro potential. This story comes to us from an article in the Visayan Daily Star in Bacolod City, dated August 31, 2000, by way of Asian Phoenix Resources.

The Problem

The Governor, Rafael Coscoluela, explained that, "the insurgents thrive in the remote areas of Negros because of lack of government presence and failure to address the basic needs of people."

Captain Cuello of the Canlusong Barangay reiterated his request for Mr. Gamboa, the Mayor of the Barangay, to look into the improvement of a farm-to-market road through rugged terrain, that had been hampering the delivery of their products to the nearby town.

Mayor Gamboa pointed out that the money for the improvements to the Canlusong Barangay road had aready been allocated by the national and provincial govenments. The problem, he suggested, was that no contractor wanted to undertake the project because of the remoteness of the place, and "the peace and order situation".

Captain Cuello admitted that there had been sightings of armed rebels in several areas. However, he maintained that the peace and order in the Barangay had remained normal with no insurgency-related incidents since the pull-out of the Army.

Even the local guerilla group, RPA-ABB, he said, had assured support for the government-initiated project in the Barangay.

The Solution

The government-initiated project that had won everyone's support turned out to be the installation of a 1,000-watt Powerpal turbine that now delivers power to a village of 23 houses and a school.

Remember, even a small turbine will power a lot of energy-efficient lights. In the mountainous rainforest, suitable sites are common.

A 1,000-watt Powerpal turbine delivers power to a village of 23 houses and a school.

Residents of the "rebel infested barangay" had the day before begun receiving electrical power from the new microhydro project.

Mayor Alfonso Gamboa had formally switched on the microhydro system in the Barangay, about 37 kilometers northwest of E. B. Magalona proper, bringing power to an initial 23 houses, including the six-room elementary school building.

Captain Cuello made the happy announcement, "At last, we have electricity," to the residents.

According to Mayor Gamboa, the 1,000-watt mini-hydro power project was a joint effort of the provincial government and the municipal government of E. B. Magalona, assisted by Representative Edith Villanueva.

Case Study 7:
A Relatively High-output System

The Problem

Although high tension lines have followed the valley bottom since their construction in the fifties, no utility power was available to the First Nations communities that lived below them. Even today this area remains remote, accessed only by a forest service road and not by the public highway system. It was in fact the routing of high tension lines through the area that brought in even this rudimentary road. BC Hydro brought access without, however, bringing electrical service. This ironical situation of millions of horsepower passing overhead while those beneath are without power is not unique. There are many places where stepping down voltage to serve a few customers is considered prohibitively expensive.

The Solution

The Peters family lives on a nice place a few miles from the village of Skookumchuck, BC, Canada. Gerard is Chief Negotiator for the In-SHUCK-ch treaty group, which represents the native people living here.

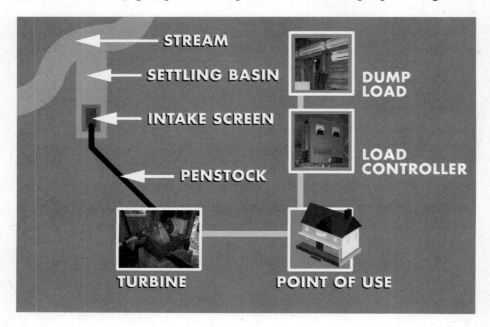

Credit:

Corri Loschuck.

When I asked, he assured me that although the name "Skookumchuck," which means "good water" in the Chinook trade jargon, didn't necessarily refer to the great hydroelectric potential in the area ... but it could. Countless streams fall thousands of feet from the snowcapped mountains to the valley floor, where most of the people live.

Steep terrain makes for lots of hydroelectric potential along the Lillooet River. Credit: Scott Davis.

Electrical service, such as it was, had been delivered by a series of diesel powered generators. It was a solution that was neither cheap nor particularly satisfactory. Like many who have spent a long time off grid, they had an unsightly (and expensive) pile of dead generators out behind the shop.

Their place, where he had built a fine log house, and was building another, actually had two small streams that were suitable for development within a few hundred feet of the house site. We chose the larger, closer one.

The site was terribly steep and rocky but it leveled out a bit at about three hundred feet above the house and was accessible by a daring logging road.

An excavator dug an intake and a trench a couple of feet deep and nine hundred feet long in a day. A crew laid and bedded the gasketted 4-inch PVC pipe. When they were done, the penstock was buried in another day's work for the excavator. The operator said, as a rule of thumb, that the machine would do 1,500 feet of ditch digging or burying per day under most conditions.

We installed a turbine alternator unit that came mounted on a skid for easy installation onto the end of the pipe. It consisted of an 8-inch Pelton wheel, direct driving a brushless 12-kilowatt alternator. It

Table 7.9: About a Relatively High Output System

Penstock length and material	4" PVC, 900 feet
Flow	Up to 350 gpm
Static Pressure	135 psi
Output	7200 kWh/month (10 kW continuous)
Turbine and Alternator	8" Pelton, 12kW brushless alternator
Consumption Rating	6
Capacity Rating	5

included a needle nozzle to provide efficient output from a variety of flows, and a husky flywheel for extra smoothness and motor starting.

We fabricated an intake from a plastic water tank by wrapping quarter-inch mesh stainless steel screen around a framework of 2-inch PVC pipe so that the water had to pass through the screen to exit the pipe. A vacuum breaker was installed.

A couple of boulders were moved in the creek so that it ponded up a bit on its plunge downward, to allow water to enter the intake box. Sandbags, used to fill the gap, can be removed to clear out accumulated gravel after a few years.

The alternator is governed by a Thompson and Howe electronic load controller. Excess load, after demand loads are met and the domestic hot water heating done, is used as space heat in winter and dumped outside to heaters on the porch in summer.

Lots and lots of electrical power. Credit: Scott Davis.

The system has run well. It provides ample power for the usual modern conveniences, plus it starts a well pump that their diesel generator would not. The system gave domestic water heating, space heat, and even a clothes dryer. A caretaker broke the needle nozzle, but it was replaced without too much difficulty. Otherwise the system has been very reliable for years.

Replacing This System Today

This system would cost less than Gerard's sport utility vehicle, approximately $15,000 (Cdn$25,000), plus labour, to replace. The turbine and alternator would be $6,000 to $7,200 (Cdn$10,000 to Cdn$12,000,) the controller $2,000 to $2,700 (Cdn$3,500 to Cdn$4,500) depending upon the options required, and the pipeline and excavator would cost about $3,900 (Cdn$6,500). Wiring and other balance of system costs would total $1,800 (Cdn$3,000). There would be taxes.

This system cost just about as much as a high quality 1-kilowatt photovoltaic system, but puts out 30 or 40 times the power. Rate this system a six in consumption and a five in capacity.

Glossary

A

AC Voltages: AC is used where voltages are higher. Our plug-ins produce 120 volts; a clothes dryer uses 240 volts.

Alternating current: The most commonly used type of electricity. It changes its direction, and so it has frequency. The graph of these changes is its wave form. Microhydro systems can generate AC directly, or obtain AC from an inverter/battery subsystem

Alternator: Generates alternating current (AC) from the rotating shaft of a turbine or fossil fuel engine.

Amp: Or ampere, is a unit of electrical current. A trickle charger at a garage charges batteries at a couple of amps. If you were comparing electricity with water, the amps would measure current flow, just as water flow is measured in gallons per minute. Amps equals volts divided by ohms. Amps equals watts divided by volts.

B

Batteries: Store electrical energy. Microhydro uses batteries so that a system need only provide the average amount of power needed, rather than having to provide for the largest anticipated loads. This makes much from little. There are different battery technologies:

- **Forklift battery**: A large and excellent deep cycle battery with a very long life expectancy. These may be more battery than is necessary for a microhydro system.
- **Golf cart battery**: A small and inexpensive kind of deep cycle battery. These are excellent for battery-based microhydro.
- **RV battery**: A compromise between an engine starting battery and a true deep cycle battery that is not recommended for use in microhydro, or in general for that matter.

Belt drive: Matches turbine speed to alternator speed. Belt drives are sometimes used in AC systems. Turbines are most efficient at certain speeds. Their rotating speed depends upon the pressure of the water. In AC systems, alternators must rotate at their rated speed. It may be necessary to use pulleys and belts get an alternator running at the correct speed when coupled to a turbine running at its most correct speed. Belt drives are not usually required in battery charging systems.

C

Capacitor start kit: Also called "easy start" kit, reduces the motor starting surge considerably.

Charge controller: Keeps batteries from overcharging.

Coanda: A screening technology that can be self-cleaning.

Conductor: A copper or aluminum wire that moves electricity from where it is generated to where it is used.

Crossflow: An impulse turbine that can be manufactured with limited technical means. See **impulse turbine**.

D

Direct current (DC): The kind of electricity batteries produce. Unlike AC, DC is steady (direct) and doesn't have frequency and waveform. Flashlights and cordless tools are common uses of DC around the home. Microhydro systems often generate DC and charge batteries.

DC charge control: Keeps batteries from overcharging in a battery charging hydroelectric system. When the battery voltage is low, current is supplied from the turbine until the batteries are fully charged. Then the charge controller turns on the diversion load until the battery voltage is low again, so that it will not be overcharged.

DC voltages: Commonly much lower than the 120/240 volt AC we use every day. For example, inverters are designed to work with battery banks that are usually 12 volts or 24 volts, and sometimes 48 volts. Some other values are used, but generally these values are lower than the 120/240 volts that is common with AC.

Deep cycle battery: A battery capable of repeated cycles of discharge and complete recharging. Engine starting batteries are actually "float" style; that is, designed to operate fully charged most of the time and to be recharged immediately. These batteries are not really designed to deep cycle.

Diversion control: Battery charging control in microhydro systems accomplished by applying a load to batteries to prevent overcharging. This feature can also be used with wind turbines. It is rarely practical to use with photovoltaics.

Diversion controller: Controls battery charging in microhydro or wind systems (rarely, photovoltaic systems as well) by applying a load to the batteries when they are fully charged.

E

Easy start kit: See **Capacitor start kit**.

F

Forebay: The structure that holds intake screening for the penstock.

Frost depth: When winters are protracted, the soil freezes. First the surface freezes, then as cold weather continues, the soil is frozen further and further down. In order to keep pipes from freezing, they would have to be buried deeper than the frost depth. Although the severity of the winter is a good indicator of frost depth, snow does insulate. A resort in the alpine zone of the Monashees (see *Home Power* magazine, #33, February/March 1999), could bury their penstock by hand to minimize environmental impact, because the soil never freezes underneath all that snow. By contrast, the case study "A Small AC System" was located in a place with much milder winters than the alpine Monashees, but because there was little snow, the frost depth was six feet down.

Francis: A reaction turbine that is very much like a centrifugal pump. See **reaction turbine**.

G

Generator: Generates DC from a rotating shaft.

Governing: Keeps AC turbine/alternator subsystems rotating at their rated speed as electrical loads change.

H

Head: The vertical distance in feet from the intake of the penstock to the turbine, measured in feet of drop. Head in feet divided by 2.31 equals pressure in pounds per square inch. Turbines are often described by the feet of head that they use, rather than the pressure required.

High head: For battery charging turbines, high heads range from over 6 feet up to 600 feet.

High pressure: May indicate a plugged nozzle.

Hybrid system: Combines two or more technologies to provide service. A photovoltaic/microhydro hybrid system is very practical where water sources dry up in the summer.

I

Impulse turbine: A turbine, such as the Turgo or Pelton, where water strikes the runner under pressure from nozzles. This is in contrast with a reaction turbine, like the Francis or the propeller type, where water is in full hydraulic contact with the runner.

Induction motor: The most common small motor, the induction motor can be used as an extremely inexpensive brushless alternator. There are efficiency and control issues, however.

Intake: Where water gets into the penstock from its source.

Inverter: Takes DC (battery power) and turns it into AC, the kind of power that comes out of household plugins.

J

Jet: The water shot at an impulse turbine runner by a nozzle.

K

Kilowatt (kW): A unit of power equal to 1000 watts. See **watt, kilowatt hour**.)

Kilowatt hour (kWh): A unit of power consumption, used for example by utilities to bill for electrical usage. One kilowatt hour is equivalent to the power consumption of 1 kilowatt in one hour. For example, if you used a 100 watt light bulb for 10 hours, you would have used 1 kilowatt hour. (See **watt.**)

Kilowatt hour per month (kWh/m): The unit that utilities use to charge for their power. It is the amount of kilowatt hours used in a month.

L

Load control: In AC governing terminology, governing is achieved through diverting loads according to priority, so that some load is on at all times. In DC charge controlling terminology, load control means somewhat the opposite. Loads are turned off, for example, when battery voltage is low. (See **Diversion control**.)

Low head: For battery charging systems, below 10 feet or so. There is some overlap with high head machines and thus room for choice.

Low pressure: May indicate low water. What is happening is that more water is exiting the system than is entering it. As a result, the pipe is no longer full.

M

Manifold: The plumbing necessary to get water to more than one nozzle.

Microhydro electricity: Generates electricity from small water-powered alternators.

N

Needle nozzle: Or variable nozzle, adds adjustability to a single nozzle by moving a specially shaped bulb in and out to vary the cross-sectional area, and thus the volume that flows through it.

Net head: Is the head left when frictional losses are subtracted from the static head. Frictional losses reduce the pressure available at the bottom of the pipe, and act as if lowering the level of the water above a gauge. Net head is found by calculating losses due to friction at a given flow rate, and subtracting them from the static head. The optimum net head in a given pipe is about two thirds of the static head.

Net pressure: Net pressure is the water pressure left when water flows through the system. This will be less than the static pressure because the friction from the water flowing through the pipes causes some pressure loss. (See **static pressure**.)

Nozzle: The nozzle shoots water in a jet at an impulse turbine. Multiple nozzles of different sizes can provide high efficiency at various flows by mixing sizes of nozzles and turning them on and off as required.

O

Off grid: Outside of the utility service area.

Ohm: A unit of electrical resistance. Ohms equals volts divided by amps. If you keep your sense of humor about comparing electricity and water, electrical resistance is like pipe resistance.

P

Pelton: The original impulse turbine, developed in the late 19th century.

Penstock: The pipe that brings water to the turbine.

Photovoltaic: System which generates power directly from sunlight (also called PV). These systems are also called "solar," but this usage can cause confusion between solar electric and solar thermal technologies.

Poly (polyethylene): An excellent pipe for microhydro in diameters up to two inches. It comes in rolls of 100 feet or more that are about five feet in diameter.

Pressure: Force exerted on an area by the water above it. The unit is pounds per square inch.

Pressure formula: Pressure in pounds per square inch $\times$ 2.31 = head in feet.

Pressure gauge: Measures pressure. An essential part of every system. With a gauge you can directly measure the optimum flow rate of an existing pipe, as well as troubleshoot problems that may arise. Low pressure may indicate that more water can exit the system than is entering it. As a result, the pipe is no longer full. High pressure may indicate a plugged nozzle.

Propeller type turbine: A reaction turbine that operates at low heads. An example is the LH-1000.

psi: Pounds per square inch.

PVC (polyvinyl chloride): A common pipe material. It is rigid, light, tough and readily available. It comes in 10- and 20-foot pieces, and in a range of pressures. The lower the pressure rating, the cheaper the pipe. Even pipe rated only to 63 psi is quite rugged, however.

R

Reaction turbine: Water is in full hydraulic contact with the turbine. Examples of reaction turbines are the low head propeller types used in the LH-1000; and the Francis turbine, which is commonly used in larger hydroelectric applications.

Rectifier: Changes AC to DC by causing to current to flow in only one direction.

RPM: Revolutions per minute.

S

Single phase AC: The most common type of electricity, using two conductors.

Small hydro: A category of waterpower that is sometimes confused with microhydro. It refers to sites that are the next size larger than microhydro, over 100 kilowatts up to a megawatt or so in size. Small hydro refers to sites that would power a town or city, and there are indeed many of these as well. However, the issues change as outputs get below a certain size, and so confusion can result. As microhydro projects get above a 100 kilowatts or so, they become small hydro projects.

Static head: The vertical distance from the turbine to the water level of the intake. Turbines are commonly rated by this factor. For example, the LH-1000 turbine requires ten feet of head.

Static pressure: The water pressure in a system when the water isn't running. This is different from net pressure, which is the pressure remaining once the water is flowing and some pressure is lost to friction. Note that the static pressure is produced by the head (sometimes called static head), the vertical distance from the intake to the turbine. Static pressure in psi = head in feet / 2.31.

Stilling basin: The first part of the intake where water is taken from its source. The water is slowed and directed into the forebay. It encourages heavy debris to stop moving, and lighter debris to float off.

Stream profile: A profile or diagram of a stream that maps survey data to show the vertical and horizontal distances. A stream profile is essential for locating the penstock in the best spot.

T

Tailrace: Where the water comes out of the turbine. (See **tailwater**.)

Tailwater: The water that comes out of the turbine, and is carried away. (See **tailrace**.)

Three phase AC: A form of AC that is very efficient to generate. Three phase AC uses three conductors compared to the two conductors used in your house to move single phase AC. Three phase AC at high voltage is used to generate and transmit microhydro power a long distance.

Transformer: Changes AC from one voltage to another to reduce transmission losses.

Triac: An electronic switching device similar in action to a dimmer, that is used in electronic load controlling technology for fine control of frequency.

Turbine: Turns a shaft by water power. The turbine runs an alternator or generator to produce power.

Turgo: An impulse turbine in common use. It can use larger nozzles than the Pelton. It was developed in the 1920s. (See **impulse turbine**.)

V

Vacuum breaker: Lets air into the pipe in case of an emergency. This is necessary because if by accident all the water rushes out from the bottom of a penstock, the resultant vacuum can collapse the pipe. A vacuum breaker lets air in at the top of the pipe to prevent damage.

Valve: Turns water on and off.

Variable nozzle: A nozzle that has a moveable bulb inside to vary the size of the jet. It provides efficient, very close control of the flow of water through the turbine over a range of flows.

Volt: A unit of electrical potential. Cars use 12-volt systems. In Canada and the US, 120-volt is standard for AC plug-ins. Volts equals watts divided by amps, or amps times ohms.

Voltage loss: Losses due to electrical resistance to current flow. If we are comparing water and electricity, voltage loss is like head or pressure loss.

W

Water level: Surveys the vertical distance in a flow of water by filling a pipe with water. A pipe, filled with water, will cease flowing when both ends are level. Bring the lower end up level with the higher end, and measure its height above the ground. This is an ancient and effective way to survey.

Watt: A unit of electrical power. Watts equal volts times amps. An energy efficient light bulb may use 16 watts to 25 watts, compared with incandescents that use 60 or 100 watts.

Weir: A device that blocks up a creek to gauge water flow by measuring the depth that the water flows over a notch or weir of a known shape in its top.

Resources

Organizations and Websites

Appropriate Energy Systems
Bob Mathews
Box 1270, Chase, BC V0E 1M0
(250) 320•8778
A great place to get microhydro! Tell him I sent you. He also gives an excellent in-person microhydro course.

Solar Energy International
www.solarenergy.org/
Offers a great microhydro course in the US.

Moorehead Valley Hydro
www.smallhydropower.com
Website for Thompson and Howe Energy Systems; features interesting case studies.

Otherpower
www.Otherpower.com
Homemade stuff, including microhydro turbines, of all sorts at this website.

Microhydro Web Portal
www.microhydropower.net
Good microhydro Web portal.

Microhydro Discussion Group
http://groups.yahoo.com/group/microhydro/

Online Guide to Microhydro Businesses
http://energy.sourceguides.com/businesses/byP/hydro/mHG/mHG.shtml

General Microhydro Information
www.picoturbine.com

Suppliers

Energy Systems and Design
P.O. Box 4557
Sussex, NB Canada E4E 5L7
Phone: (506) 433•3151
Fax: (506) 433•6151
E-mail: hydropow@nbnet.nb.ca
Website: www.microhydropower.com/
Manufacturer of Stream Engine, LH-1000 and
Water Baby Turbines.

Asian Phoenix Resources Ltd.
2 – 416 Dallas Road
Victoria, BC Canada V8V 1A9
Phone: (250) 361•4348
Fax: (250) 360•9012
E-mail: info@powerpal.com
Website: www.powerpal.com
Manufacturer and distributor of microhydro equipment,
including the Powerpal.

Harris Hydroelectric
632 Swanton Road
Davenport, CA 95017
Phone: (831) 425•7652
E-mail: harrishydro@cruzers.com
Manufacturer of Harris Pelton turbine.

Nautilus Water Turbine
2131 Harmonyville Road
Pottstown, PA 19465
Tel: (610) 469•1858
Fax: (610) 469•1859
Website: www.waterturbine.com/
Manufacturer of Nautilus low head turbines.

Soltek Solar Energy Ltd.
2 – 745 Vanalman
Victoria BC v8z 3B6
(800) 667•6527
Website: www.spsenergy.com
Canadian supplier of Platypus turbines from Australia.

Thomson and Howe Energy Systems Inc.
8107 Highway 95A
Kimberly, British Columbia, Canada v1A 3L6
Phone: (250) 427•4326
Fax: (250) 427•3577
E-mail: thes@cyberlink.bc.ca
Manufacturer of electronic load controllers and contact for
pumps-as-turbines.

Canyon Industries Inc.
5500 Blue Heron Lane
Deming, WA. 98244 USA
Phone: (360) 592•5552
Fax: (360) 592•2235

E-mail: CITurbine@aol.com
Website: www.canyonindustriesinc.com/
Turbine manufacturer.

Dependable Turbines Ltd.
17930 Roan Place
Surrey, BC Canada v3s 5k1
Phone: (604) 576•3175
Fax: (604) 576•3183
E-mail: sales@dtlhydro.com
Website: www.dtlhydro.com/
Turbine manufacturer.

Real Goods
Phone: (800) 762•7325
Website: www.realgoods.com/
Great place to get all sorts of equipment, including turbines.

Publications and Videos

Home Power **Magazine**
P.O. Box 520
Ashland, OR 97520 USA
Website: www.homepower.com
Read all about it! An excellent source for all things renewable.

Residential Microhydro with Don Harris, 45 minutes, 1997
A video about microhydro.

Waterpower, 25 minutes, 1981
A Canadian video about microhydro.

Stand-Alone Micro-hydropower Systems: A Buyer's Guide Ottawa, 2003.
Natural Resources Canada, Renewable and Electrical Energy Division,

Energy Resources Branch.
Available soon for download at the Canadian Renewable Energy Network.
Website: www.canren.gc.ca.
Available in future in paper format as well.

Going with the Flow: Small Scale Water Power
Dan Curtis
UK-based microhydro book with more sample microhydro systems.
Available from Centre for Alternative Technology Publications.
Website: www.cat.org.uk

Index

B

batteries, 2-3, 9, 55, 65, 69, 74-6, 82, 112, 131
    charging, 15, 53, 67-8, 79, 103-4, 108
    deep cycle, 66, 132-3
    forklift, 66, 110, 132
    golf-cart, 54-5, 66, 106, 110, 132
    overcharging, 132
    RV type, 66, 132
    see also battery charging systems

battery charging systems, 2, 7, 23-4, 28, 53, 55-6, 61-2, 68-9, 74, 76, 78-9, 82, 122

battery storage, 27, 54-5, 106

battery subsystems, 8, 15, 20, 28, 54, 74, 106, 110, 113, 131

BC Hydro, 127

Buck Creek Ranch (BC), 24, 57, 76, 78, 88, 114-5, 117-8, 120-4

C

Canadian Renewable Energy Guide (1999), 78

capacitor start kit, 132

capacity, power, 12, 24, 26

Capacity Self-rating Scale, 22, 24-5, 68, 74, 105, 108, 111-2, 115-6, 121-3, 129-30

centrifugal pump, 64

charge controller, 55, 67, 132-3

Coanda effect, 97, 132

cogeneration, 100

conductor, electrical, 5, 132

conservation, energy, *see* energy conservation

conservation of fossil fuels, 16

consumption, fuel, 16

G

gel cell, 67

generators, 16, 20, 27, 75, 106-7, 128-9, 134
    governing, 82, 115, 118, 129, 134
    technology for, 71, 73

gravity water system, 26, 85

H

Harris Hydroelectric
    alternator, 59-60
    Pelton impulse turbine, 58-9
    turbine, 105-6
    see also alternators; turbines

head loss, 5, 45-7, 50-1, 84

heating, resistance, 56

heating, solar, 23-4

high head systems, 34, 38, 81, 89
    see also low head systems

hose bib, 36, 82, 105

hydro projects, small, 138

hydroelectric generation, 101

hydroelectric plants, 71

hydroelectric systems, *see* microhydroelectric systems

hydroelectricity, 1, 19, 88

I

impulse wheels, *see* turbines, impulse

induction motors, 57, 65, 135

intake, screening methods, 96-7, 101-2, 133, 135

intake system, 81, 86-7, 89-91, 93-6, 102, 105, 108, 123-4, 128-9, 139

inverter, sine wave, 25, 68-9, 73-4, 109-10

PV-powered systems, *see* photovoltaic systems

R

recovery time, hot water tank, 23, 75

relay, solid state, 73, 118

remote power systems, 119

renewable technologies, 1, 11

resistance, electrical, 5-6

resistance heating, 56

RETCAP program, 100

S

sine wave AC waveform, 62, 74

site assessment, 2, 10, 16, 53

small hydro projects, 138

solar energy systems, 10-1, 13, 104
    see also photovoltaic systems

solar heat, *see* heating, solar

static head, 5, 43-6, 50-2, 109, 111-2, 122, 138

stilling basin, 88, 95, 139

stream profile, 40, 42, 53, 139

Sunfrost refrigerator, 22

surge capacity, 56, 74

T

tailrace, 139

tailwater, 139

technology, renewable energy, 1, 11

Tesla, Nikola, 6

Thompson and Howe electronic load controller, 76, 129

About the Author

SCOTT DAVIS IS AN AWARD-WINNING renewable energy project developer who has built, designed, operated, sold, repaired, and generally fooled around with many successful water, wind and solar power systems. He is currently marketing his latest project, *Microhydro from Water to Wire*, a 22-minute online movie, as well as working on his memoirs.

If you have enjoyed *Microhydro,* you might also enjoy other

BOOKS TO BUILD A NEW SOCIETY

Our books provide positive solutions for people who want to make a difference. We specialize in:

Sustainable Living • Ecological Design and Planning

Natural Building & Appropriate Technology • New Forestry

Environment and Justice • Conscientious Commerce

Progressive Leadership • Resistance and Community • Nonviolence

Educational and Parenting Resources

New Society Publishers

ENVIRONMENTAL BENEFITS STATEMENT

New Society Publishers has chosen to produce this book on New Leaf EcoBook 100, recycled paper made with 100% post consumer waste, processed chlorine free, and old growth free.

For every 5,000 books printed, New Society saves the following resources:[1]

29	Trees
2,609	Pounds of Solid Waste
2,870	Gallons of Water
3,744	Kilowatt Hours of Electricity
4,742	Pounds of Greenhouse Gases
20	Pounds of HAPs, VOCs, and AOX Combined
7	Cubic Yards of Landfill Space

---

[1]Environmental benefits are calculated based on research done by the Environmental Defense Fund and other members of the Paper Task Force who study the environmental impacts of the paper industry.

For more information on this environmental benefits statement, or to inquire about environmentally friendly papers, please contact New Leaf Paper – info@newleafpaper.com Tel: 888 • 989 • 5323.

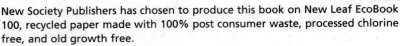

For a full list of NSP's titles, please call 1-800-567-6772 *or check out our web site at:*

www.newsociety.com

NEW SOCIETY PUBLISHERS